Transformations of Warfare in the Contemporary World

In the series **Politics, History, and Social Change**,
edited by John C. Torpey

Heribert Adam and Kogila Moodley, *Imagined Liberation: Xenophobia, Citizenship, and Identity in South Africa, Germany, and Canada*

Aidan McGarry and James Jasper, *The Identity Dilemma: Social Movements and Collective Identity*

Philipp H. Lepenies, *Art, Politics, and Development: How Linear Perspective Shaped Policies in the Western World*

Andrei S. Markovits and Emily Albertson, *Sportista: Female Fandom in the United States*

Nicholas Toloudis, *Teaching Marianne and Uncle Sam: Public Education, State Centralization, and Teacher Unionism in France and the United States*

Philip S. Gorski, *The Protestant Ethic Revisited*

Étienne Balibar, Sandro Mezzadra, and Ranabir Samaddar, eds., *The Borders of Justice*

Kenneth H. Tucker, Jr., *Workers of the World, Enjoy! Aesthetic Politics from Revolutionary Syndicalism to the Global Justice Movement*

Hans-Lukas Kieser, *Nearest East: American Millennialism and Mission to the Middle East*

Ernesto Verdeja, *Unchopping a Tree: Reconciliation in the Aftermath of Political Violence*

Rebecca Jean Emigh, *The Undevelopment of Capitalism: Sectors and Markets in Fifteenth-Century Tuscany*

Aristide R. Zolberg, *How Many Exceptionalisms? Explorations in Comparative Macroanalysis*

Thomas Brudholm, *Resentment's Virtue: Jean Améry and the Refusal to Forgive*

Patricia Hill Collins, *From Black Power to Hip Hop: Racism, Nationalism, and Feminism*

Daniel Levy and Natan Sznaider, translated by Assenka Oksiloff, *The Holocaust and Memory in the Global Age*

Brian A. Weiner, *Sins of the Parents: The Politics of National Apologies in the United States*

Heribert Adam and Kogila Moodley, *Seeking Mandela: Peacemaking Between Israelis and Palestinians*

Transformations of Warfare in the Contemporary World

Edited by John Torpey and David Jacobson

TEMPLE UNIVERSITY PRESS
Philadelphia • Rome • Tokyo

TEMPLE UNIVERSITY PRESS
Philadelphia, Pennsylvania 19122
www.temple.edu/tempress

Published 2016

Library of Congress Cataloging-in-Publication Data

Names: Torpey, John, 1959– editor. | Jacobson, David, 1959– editor.
Title: Transformations of warfare in the contemporary world / edited by John Torpey and David Jacobson.
Description: Philadelphia : Temple University Press, [2016] | Series: Politics, history, and social change | Includes index.
Identifiers: LCCN 2015044971| ISBN 9781439913123 (cloth : alk. paper) | ISBN 9781439913130 (paper : alk. paper) | ISBN 9781439913147 (e-book)
Subjects: LCSH: War—History—21st century. | Military art and science—United States.
Classification: LCC U21.2 .T695 2016 | DDC 355.02—dc23 LC record available at http://lccn.loc.gov/2015044971

∞ The paper used in this publication meets the requirements of the American National Standard for Information Sciences—Permanence of Paper for Printed Library Materials, ANSI Z39.48-1992

Printed in the United States of America

9 8 7 6 5 4 3 2 1

Contents

Preface and Acknowledgments

On March 21, 2015, the *New York Times* reported that "the Islamic State has called on its members and sympathizers in the United States to kill 100 service members whose names, photos and purported addresses it posted on a website. The group said that the personnel had participated in efforts to defeat it in Syria, Iraq, Yemen and elsewhere."[1] There could hardly be a better example of how warfare has been transformed in recent years. Far from the sorts of wars that once involved massive armies squaring off against one another on, say, the plains of Central Europe, much conflict today concerns relatively small numbers of antagonists engaging in precision attacks using highly sophisticated personnel and technological tools.

If what the Islamic State wants to do here can be said to be warfare at all rather than simply terrorism, it is engaged in targeting very specific targets—all the way down to the level of individuals. From this perspective, the attack on the World Trade Center towers and the Pentagon in 2001 was unusual, targeting as it did a relatively large number of humans; but the point of those attacks was the symbolic rather than the strategic significance of the deaths that were caused and the buildings attacked. In response, the United States ramped up a program of drone attacks on suspected terrorists in the Middle East, Pakistan, and elsewhere, which themselves targeted specific individuals held to be implicated in terrorist acts.

It is therefore ironic that, when U.S. president Barack Obama spoke about the nature of possible U.S. military intervention in Syria after President Bashar al-Assad used chemical weapons on his population in the summer of 2013, Obama insisted, "The United States military doesn't do pinpricks."[2] Given the transformation of warfare in the contemporary period, that is, in fact, precisely what the U.S. military does do more and more. Conflict involving countries in the core of the world system has shifted away from warfare between massed national populations and toward fighting involving small numbers of combatants using high-tech weaponry of various kinds—drones, robots, cyber-weapons, sophisticated special operations forces, and the like. This shift is why the Russian incursion into Crimea and Ukraine in early 2014 elicited so many comments about how anachronistic it seemed.

Since 1945, Europe—or at least Western Europe—has been utterly transformed by the pacific character of its international relations, leading one prominent historian to ask, in the title of a valuable book on the subject, "Where have all the soldiers gone?"[3] One answer is that they have been "civilianized," as participation in military service in Western Europe and the United States has become a comparatively marginal activity and the functions associated with warfare are increasingly attended to by voluntary specialists who choose their work more as a career decision than as the fulfillment of a sacred duty. In the poorer parts of the world, alas, soldiering involves both masses of the population, though often as innocent bystanders, and specially committed volunteers (particularly in the case of insurgents). The bright side of this picture is that much warfare now is internal rather than international and the kinds of weaponry involved can do relatively little damage when compared to the aerial bombings, heavy artillery, and nuclear weapons that devastated World War II–era Europe and Japan. In view of the deadly conflicts and atrocities that continue to plague the contemporary world, Steven Pinker's much-noted analysis (building on the sociologist Norbert Elias's seminal work) of the declining risk of violent death may be overly rosy, but his study accurately captures the major trend in recent years: military violence and the risk of violent death have been reduced by comparison with earlier centuries, and the trend promises to continue as a result of transformations in the character of war in our day. As Robert Jay Lifton has pointed out, nuclear weapons could change this calculus very quickly, but they may also be responsible for the decline in conflict among the great powers.[4] The risks are too great to contemplate.

This book addresses some of the dramatic changes that have been taking place in the realm of warfare in recent years. It discusses the waning connection for most people between citizenship and soldiering, the shift toward more reconstructive than destructive activities by militaries, the ethics of irregular, or asymmetrical, warfare, the role of novel techniques of identification in military settings, the stress on precision associated with targeted killings and kidnappings, and the uses of the social sciences in contemporary warfare. Two of the chapters, in particular, are skeptical that much has really changed; Rob Johnson and Roberto González suggest that, changes in technology and language notwithstanding, the basic nature of warfare remains unchanged, and social scientific embellishments do little to affect that nature.

In his concluding remarks on the book, David Jacobson explores the extent to which the contemporary transformation of warfare is a product of a shift in the character of our enemies—from other nation-states to postnational groups, especially Islamic extremists who wish to re-create a caliphate governing all Muslims in which non-Muslims would be either *dhimmi* ("protected persons" without certain rights and obligations associated with Muslims) or banished altogether. The turn toward precision attacks, away from citizen-soldiering, and against the nation-state do seem to go hand in hand. The transformation of warfare is thus not driven by any single factor; instead, it reflects changes in a variety of technological, strategic, ideological, and ethical dimensions.

This book is offered as a contribution to a discussion of these issues, which force us to try to make sense of the shifting but possibly less dangerous terrain of military conflict in the years to come. It may be too much to characterize contemporary warfare as smart warfare, but many of its features suggest that—while all war may be dumb in the sense that we have been unable to find less harmful ways to resolve differences—contemporary warfare may be conducted in more sophisticated and numerically less deadly ways than in the past. Insofar as that is the case, it would be a good thing for the human species, which often seems intent on destroying itself in one way or another.

We thank a number of institutions and individuals for their help in making this book possible. First, we are grateful to a number of supporters at the Graduate Center of the City University of New York: the Ph.D. Program in Sociology; the Advanced Research Collaborative and its director, Professor Don Robotham; the Ralph Bunche Institute for International Studies (of which John Torpey is now director); and the Office of the Provost for its assistance in organizing a workshop in early 2013 where some of these ideas were first aired. John Torpey thanks Saskia Hooiveld for her assistance with the workshop and her subsequent involvement in the writing of Chapter 1.

We are also grateful for the support of the Global Initiative on Civil Society and Conflict at the University of South Florida and to Derek Harvey for codirecting and arranging a broad array of academic, government, and independent sector participants for the conference "Transformation of War," held in Tampa in early 2014.

Finally, we are grateful to Danielle Zach, now at City College, for her help in making the final text more readable.

Notes

1. Michael S. Schmidt and Helene Cooper, "ISIS Urges Sympathizers to Kill U.S. Service Members It Identifies on Website," *New York Times*, March 21, 2015, available at http://www.nytimes.com/2015/03/22/world/middleeast/isis-urges-sympathizers-to-kill-us-service-members-it-identifies-on-website.html.

2. White House, "Remarks by the President in Address to the Nation on Syria," September 10, 2013, available at https://www.whitehouse.gov/the-press-office/2013/09/10/remarks-president-address-nation-syria.

3. James J. Sheehan, *Where Have All the Soldiers Gone? The Transformation of Modern Europe* (Boston: Houghton Mifflin, 2008).

4. See the debate between Pinker and Lifton at the New School on March 23, 2012, available at https://www.youtube.com/watch?v=m1-kXmCgWT0.

Abbreviations

ABIS	Automated Biometric Identification System
AFIS	Automated Fingerprint Identification System
AOR	Area of Responsibility
BATS	Biometric Automated Toolset
BIMA	Biometrics Identity Management Agency
BISA	Biometric Identification System for Access
BMO	Biometrics Management Office
CIA	Central Intelligence Agency
COIN	counterinsurgency
DHS	Department of Homeland Security
DoD	Department of Defense
EOD	explosive ordinance disposal
FBI	Federal Bureau of Investigation
HIIDE	Handheld Interagency Identity Detection Equipment
HTS	Human Terrain System
HTT	human terrain team
IDF	Israeli Defense Force
IED	improvised explosive device
ISF	Iraqi Security Forces
ISIL	Islamic State of Iraq and the Levant

NATO	North Atlantic Treaty Organization
NTC	National Training Center
PMSC	private military and security contractor
RMA	revolution in military affairs
W-ICEWS	World-Wide Integrated Crisis Early Warning System

1 / Warfare without Warriors?

Changes in Contemporary Warfare and the Demise of the Citizen-Soldier

John Torpey and
Saskia Hooiveld

We appear to be in the midst of a momentous transformation in human affairs. The period when mass armies fought over territories to control their population and resources is fading in the developed world and becoming a phenomenon of the world's poorer precincts. Fighting at the behest of the world's wealthiest countries is shifting dramatically toward forms that require few citizen-soldiers—what one might call warfare without warriors. At least in the rich world, combat between mass armies is being replaced by cyber-warfare, pilotless aircraft, special operations commandos in out-of-the-way hot spots, and even robots as the predominant forms of military operations. *Multinational* private military and security contractors (PMSCs) conduct operations in support of *national* security interests. Battlefields are harder to discern, and the line between warfare and police work has grown murky. The world's affluent countries have become increasingly unwilling to involve their own armies in combat, and warfare has become a scourge concentrated geographically in what one might call the global ghetto.

Consider the distance that has been traveled from the Battle of the Somme one hundred years ago, when aristocratic officers led a generation of young men to its death for a few miles of land in the heart of Europe. World War I took place at the high tide of European nationalism, and the association between citizen and state was seen as embodied in the willingness of nationals to die for their country. Indeed, the citizen-soldier has long been held to be the backbone of the modern nation-state, demonstrating in the most profound way the interlocking of warfare, state, and nation.[1] Yet recent developments that marginalize the military in the affluent world raise crucial questions: To what degree has the citizen-soldier actually been an empirical historical reality? And will the citizen-soldier endure given the changing character of warfare?

We argue that the veneration of the citizen-soldier ideal reflects a historical anomaly and that the citizen-soldier phenomenon in the globe's more affluent regions is on its way out. The period during which citizenries were the chief source of the soldiery is relatively circumscribed historically, extending over a couple of hundred years at most. Some argue that we are observing a return to a pattern that existed before the rise of the nation-state, when European rulers hired military contractors to fight their battles and the notion of the citizen-soldier had little relevance. While we agree with that basic view, we argue that the disappearance of the citizen-soldier is a result of more complicated processes than merely the general trend toward the privatization of the state's functions associated with neoliberalism, including through military contracting. It is necessary to consider important technological developments and military recruitment, as well as the emergence of social norms in the world's more affluent countries that challenge the very legitimacy of war and render problematic the activity of soldiering.

At the same time, we note a countervailing trend that has revalued participation in the military: while soldiering has grown less and less significant and prestigious as an obligation of citizenship in affluent democracies, previously excluded groups—blacks, gays and lesbians, and women—have in recent decades made recognition of their participation in military activities a major aspect of their quest for full inclusion in the body politic. This in itself also reflects the "taming" of the warrior, however, and renders participation in the military an aspect of the ordinary give-and-take of democratic politics in the developed world. Meanwhile, in many countries from the poorer parts of the world, membership in the armed forces is often a major source of prestige, political advantage, and income and hence a matter over which various groups struggle for access and control. The relationship between citizenship and military service is thus complex and varied according to time and place.

In what follows, we explore the participation of soldiers in the modern nation-state and the increasing use of impersonal forms of warfare among affluent states. We survey participation in warfare, on behalf of various states and according to several possible categories of combatants—citizens, mercenaries, colonial soldiers, professional soldiers, PMSCs, and others—and how these have changed over time. We document the shift in Western European and American society from (occasional uses of) conscription to professional armies in the context of a relatively pacific period in their histories. We conclude by examining the changing nature of warfare today and its impact on the citizen-soldier ideal.

Before the Citizen-Soldier

The characteristic pattern of European warfare until at least the waning days of the eighteenth-century ancien régime, and in some contexts well into the nineteenth century, was for military affairs to be organized on the basis of private contracts, whether these were for the recruitment and maintenance of fighting soldiers, for the provision of military hardware and munitions, or for military support systems. Typically, there were "varying forms of *private-public partnership*, in which very substantial elements of private contracting, finance and administration" prevailed in systems that might still have had "a core of state-raised or state-maintained troops."[2]

Needless to say, the permanent garrisoning of troops is costly for rulers. Fiscal constraints on raising an army—the difficulties of acquiring revenue without satisfactory systems of taxation—were a major reason behind the preference for public-private partnerships. As Michael Mann has noted, "Agrarian states [that is, all states before the nineteenth century at the earliest] could not even know the worth of their subjects, let alone tax them accurately. . . . [T]oday the American and British states can both tax my own income and wealth 'at source' . . . and extract their cut without my even laying hands on it."[3] Advances in taxation under centralizing nation-states made it possible to raise revenues that could support larger military forces. "Embracing" male citizens to make them available for military service was another aspect of modern states' growing infrastructural power, the most essential enhancement of their capabilities compared to earlier states.

After the American Revolution, however, with its glorification of republican principles, the norm against using mercenaries gathered force in Europe and the United States. Consistent with republican conceptions of virtue, the notion of soldiering for hire acquired an increasingly foul odor among those imbued with the ideal of self-government, and the use of mercenaries declined. By the mid-nineteenth century, late-adopter Britain finally fell in line with the "anti-mercenary norm" that had grown in strength over the preceding decades, and mercenaries largely disappeared from battle after the Crimean War.[4] Soon thereafter, Britain adopted the Foreign Enlistment Act, which prohibited its subjects from participating in warfare on behalf of any state with which Britain itself was at peace. In the interim, many countries have adopted similar laws forbidding their citizens from going to war on behalf of another state. The 1989 International Convention against the Recruitment, Use, Financing and Training of Mercenaries defined a mercenary as a combatant recruited abroad who "is motivated . . . essentially by the desire for private gain."[5] Such motivation had come to be regarded as incompatible with the selflessness associated with patriotic duty.

Reversing the position of laws that constrain immigration while taking a generally laissez-faire view of emigration, the national laws seeking to outlaw the soldier of fortune seem mainly concerned to deter nationals from departing to fight on behalf of other nations. In contrast, foreigners wishing—or forced—to fight on a nation's behalf have been viewed considerably more charitably. It has therefore not been the case that only citizens participated in modern warfare on behalf of their own states, even in the age of the democratic nation-state inaugurated by the American and French Revolutions. This was in part because overseas colonization was often spearheaded first not by states but by joint-stock companies (authorized by European sovereigns) seeking to make a profit for themselves and their investors, such as the British East India Company and the Dutch East

India Company. At the beginning of the modern colonial period, troops in European overseas possessions were recruited predominantly from the home army of the country concerned, but locally raised native troops were soon mustered as well. The latter normally served in separate units, at first under their own leaders and later under European officers.

The sepoys of the British East India Company were a major early example. By the mid-eighteenth century, these troops were beginning to be directly recruited and officered by the company. Some of the sepoys rebelled against the company during the Sepoy Mutiny of 1857, leading to the end of the British East India Company's rule in India. After the British Raj took control in 1858, the sepoys formed the famous regiments of the British Indian Army, some of which survive to the present day in the national armies of Pakistan and India. Meanwhile, the French Foreign Legion was specifically created in 1831 to recruit foreign nationals wishing to fight on behalf of France. The Royal Netherlands Indian Army (referring to the Dutch East Indies, now Indonesia) was created in 1830, a year before the French Foreign Legion, and had a similar recruitment policy. It stopped being an army of foreigners around 1900 when recruitment was restricted to Dutch citizens and to the indigenous peoples of the Dutch East Indies.

In the larger colonial possessions, the garrison was likely to consist of both locally recruited and white European troops. The latter might be from the home or metropolitan army, from settlers doing their military service, or occasionally from mercenaries recruited outside the territories of the colonial power concerned. The Dutch had a mix of locally recruited and metropolitan troops making up their garrison in the East Indies. While the Sikhs, Punjabis, Jats, Baluchis, and other "martial races" making up the bulk of the Indian Army were recruited from British subjects, the ten regiments of Gurkha Rifles were recruited from outside British territory. Some of these groups would come to be regarded as martial races, their martial qualities propagandized, and their communities rewarded with special status. Given the proportions involved, however, the colonial power using these arrangements might find itself confronted with a dilemma: when military developments made numbers a priority, it had to either trust the majority, and so risk loss of control, or rely on minorities combined with large numbers of expensive European or other nonlocal troops.

By the twentieth century, colonial troops were often being used outside the boundaries of their territories of origin. Troops from France's North African colonies served in the trenches during World War I in France itself. Italy employed Dubats from Somaliland, together with Eritrean and Libyan units, in the conquest of Ethiopia during 1936. France also made extensive use of African troops in World War II and during the subsequent Indochina and Algerian Wars. Indian troops served in Europe in large numbers during both world wars, as well as in the Middle East, Malaya, Burma, and North Africa in World War II. The Nazi SS made extensive use of foreigners before and during World War II, and Japan recruited levies from Korea and Taiwan during colonial rule in both countries. In sum, modern nation-states putatively committed to the norm of the citizen-soldier were also happy to include foreigners among their troops when doing so served their purposes.

The Citizen-Soldier Ideal

The impact of the citizen-soldier ideal on civil-military relations in the United States has been manifest from the country's birth in 1776. The influence of the ideal is apparent in the Constitution's Second Amendment as well as in the Militia Act of 1792. The Second Amendment asserted that a "well regulated Militia" was "necessary for the security of a free state," and hence the "right of the people to keep and bear Arms, shall not be infringed." Soon after the Bill of Rights was ratified by Congress in late 1791, the Militia Act required the service of every able-bodied, white, male citizen between the ages of eighteen and forty-five in his state's militia.[6] In addition to these temporary soldiers, however, full-time troops supplemented part-time local militias. This force was staffed by citizen volunteers and subordinated to civilian control—two safeguards intended to prevent the emergence of an autonomous military class disconnected from civil society and capable of overthrowing democratic government.

Over the course of American history, the military importance of the citizens' militia eroded, yet the conceptual connection between military service and citizenship at its core persisted. As the United States moved from a confederation of individual states to become a centralized nation-state after the Civil War (1861–1865), the 1903 Militia Act put the National Guard under closer federal supervision. The Selective Service Act of 1917 authorized the president to conscript soldiers for combat in World War I, but it was rescinded in 1920 after the war ended. Selective service was reinstated in 1940 on the eve of U.S. engagement in World War II, but it ended after the conflict concluded. Drafts were reinstated for the Korean War and the Vietnam War. Despite its episodic nature in American history, compulsory military service was generally viewed as an important civic duty for men.[7] At least until Bill Clinton was elected president in 1992, military service was regarded as a crucial requirement for high public office in the United States. That norm began to subside in subsequent years, in tandem with the military's decline in significance in American life more generally.

In Europe, meanwhile, militaries recruited by conscription up to World War I inculcated in the broader society the virtues of discipline, patriotism, and self-sacrifice.[8] Peasants became Frenchmen by attending schools and through their participation in the armed forces, just as armies are today often the backbone of common nationhood. Even in postrevolutionary France, armies were led by officer corps drawn heavily from among the aristocracy, and military service was frequently portrayed as gallant and noble; this was a prominent element of what Arno Mayer calls the "persistence of the old regime."[9] However much military service might have been glamorized for the officer corps, ordinary soldiers were typically those unlucky enough to not escape the military recruiter or too poor to purchase substitutes. The martial virtues were on display in such influential literary works as Ernst Jünger's *Storm of Steel*.[10]

The mass death occasioned by World War I helped undermine the enthusiasm for armed combat, as evidenced in Robert Graves's 1929 memoir *Good-bye to All That* and Sigmund Freud's discovery of a "death-instinct."[11] The cataclysm of World War II further took the bloom off the militarist rose and was followed by the so-called long peace that saw the geographic boundaries of Europe frozen in place on either side of the Iron Curtain. During that forty-year period, Europeans also grew more devoted to peace and human rights.

The unraveling of Yugoslavia and the attendant warfare, especially in Bosnia, came as a shock to what had become, by comparison to European sensibilities before the end of World War II, a peace-loving conscience. Indeed, wars among the great powers have become seemingly a thing of the past, and interstate wars—the kind that arrayed the massed citizens of one country against those of another—declined in frequency, while the world witnessed the relative rise of intrastate wars, almost exclusively in the global South.[12]

Professional Forces and the Fate of the Citizen-Soldier

Attitudes concerning military force changed dramatically in the post–World War II period in Western Europe and North America. The mass army based on conscription with extensive reserves was gradually phased out of existence in Western countries. Such armies were replaced by professional or volunteer armed forces and by militaries in which conscripts chose military duty rather than readily available alternative-service positions, such as in Austria and Germany, where they typically staffed group residential facilities, performed varieties of social work, and the like. Britain adopted a professional army in 1960. The United States shifted to an all-volunteer force in 1973 as the Vietnam War wound down amid much protest and social conflict. The end of the Cold War brought a fresh movement toward professionalized military forces. The Dutch ended compulsory military service in the 1990s. France abandoned conscription in 2001. Germany eliminated the draft in 2011. In an apparent exception to the trend, Austrians voted in early 2013 to maintain conscription, but in reality they did so mainly to ensure that they would continue to have the large pool of cheap labor that performs civil service (*Zivildienst*) in the country's old-age homes, hospitals, and youth centers.[13]

These professional forces were quite different from conscripted military forces. After 1945, Western European armies were increasingly influenced by the values and habits of civilian institutions.[14] Accordingly, the expectation that the leaders of Western nation-states could call on citizens to make the greatest sacrifice was in decline. David Segal concludes that military service in the United States has been redefined "from being an obligation of citizenship in a community to being an obligation of national citizenship and, most recently, to being a job."[15] In the early twenty-first century, compulsory military service exists in only a relatively small number of countries in the world.

Many argue that the elimination of the draft in North America and Europe has severed the link between military service and citizenship. Elliot Abrams and Andrew Bacevich assert that "the mythic tradition of the citizen-soldier is dead." Similarly Daniel Moran notes that "the legend of the *levée en masse* has, to all appearances, lost its grip upon the Western imagination." James Burk writes that, with the end of the draft, the United States "abandon[ed] the ideal of the citizen soldier, conscripted into the mass army."[16] For these authors, the demise of the citizen-soldier tradition is also associated with a host of ills: a corrosive and conflict-ridden culture of rights, national disunity, and an unstable social system in which the burden of defense is not borne equally by all segments of the society.

But these conclusions may not follow if the premise is faulty. For example, with regard to the American case, Ronald Krebs argues that the role of the citizen-soldier as the backbone of the U.S. military has been exaggerated and therefore should not be used as a baseline to assess the place of the military in American life.[17] Indeed, the lionization of the so-called greatest generation, which fought in World War II, has distorted people's perceptions of the importance of the citizen-soldier in American life. Krebs also objects to the expectation in the republican tradition that citizens must put their lives on the line to merit full membership in the polity.[18] Given that such service has historically been largely restricted to men, his objection has considerable force. In all events, recent developments in the nature of warfare have made the citizen-soldier seem increasingly irrelevant to the tasks of war making and security.

Warfare without Warriors, and Remote-Control Dominance

Since the end of World War II, the world's wealthier countries have become increasingly unwilling to involve their own armies in combat. Indeed, a repeat of that massive conflagration involving major Western European states borders on the inconceivable. In part because of the overwhelming significance of nuclear weapons, the prospect of mass armies of conscripts fighting over pieces of European territory has faded in importance; as a security matter, Europeans today are mainly concerned about Islamist terrorists, not enemy states. In short, twenty-first-century warfare looks very different from that which characterized the period before 1945 or even 1989, when the Cold War pitted the massed armies and nuclear stockpiles of the Soviet bloc and the North Atlantic Treaty Organization (NATO), led by the United States.

In contrast to their earlier reliance on mass armies of conscripts, affluent countries now tend to rely on enlisted forces drawn from among the poorer and less-educated segments of their societies. The size of the armed forces of wealthy countries has generally declined in recent years.[19] The ranks of the U.S. military have decreased since the spike of the Vietnam War years and the shift to a nonconscript force; the Iraq and Afghanistan wars were fought with roughly the same number of soldiers as were under arms at the end of the Clinton administration, which had sharply cut military budgets as a proportion of gross domestic product.[20] Karl Marx once noted that, ironically, in contrast to the military, which was stronger if it had more men under arms, the goal of the capitalist was to eliminate workers from the industrial army. Increasingly in the wealthy world, however, the military has fallen in line with the pattern Marx associated with capitalist enterprises: the number of people under arms steadily shrinks as a proportion of population, mirroring the downsizing of the industrial labor force.

The reductions in combat personnel are in part a product of budget cuts, in part a result of shifting patterns of war fighting. The citizen-soldier of yore is increasingly being replaced by advanced technology, elite special forces, or PMSCs of indeterminate nationality. More than half the personnel deployed by the United States in Iraq and Afghanistan since 2003 were employees of PMSCs—modern-day mercenaries, raised by perfectly legal contracting businesses rather than by the enterprising warlords one finds in other settings.[21] In the process, the U.S. military looks more and more like the mixed forces of regulars and mercenaries that fought for multinational empires of the past and before the establishment of the norm against mercenaries in the nineteenth century.

Special operations forces—small numbers of highly trained and expensively outfitted troops—are surgically inserted into specialized capture-or-kill missions rather than deployed to slog it out with the enemy over an extended period. The norm of the citizen-soldier as the source of military protection in nation-states—insofar as it ever existed—has thus been sharply attenuated and indeed is growing obsolete. From the point of view of the average soldier, the armed forces of rich countries have shifted away from destroying and toward building—schools, hospitals, nations. Their activities have often become as much humanitarian or quasi-diplomatic as military. Meanwhile, counterinsurgency tactics are deployed simultaneously in Central Asia and East Harlem.[22] Armed forces still kill other people, of course, but their activities have grown rather more diverse than that age-old endeavor.

This is to a considerable extent because most warfare today takes the form of civil wars and insurgencies in the globe's poorer parts, and the building activities are intended to stem instability and terrorism. These conflicts involve by definition not only a country's official armed forces—more or less well trained and equipped—but a variety of other combatants as well. The worldwide recruitment of the soldiery from among the poor is further evident in United Nations–sponsored peacekeeping missions, in which the blue helmets are drawn predominantly from the populations of an array of countries in the global South and their service is paid for by the wealthy benefactors of the UN. Relying on the poor and disenfranchised may make it easier for countries to commit themselves to military conflict, knowing that the children of elites will be shielded from the consequences of their decisions. This was, of course, one of the main arguments in favor of the citizen-soldier model.

Yet with changes in the very conduct of warfare, soldiers from the global North are less and less likely to be sent into battle in the first place. The operators of unmanned aerial vehicles, better known as drones, can strike a supposed enemy combatant from thousands of miles away. These warriors by remote control may do their jobs from nine to five and then leave for a Parent-Teacher Association meeting at the end of the day. That drone technology has been used to kill a small number of U.S. citizens has been cause for special concern among the American public. These citizens were not necessarily charged with any crime, not about to perpetrate an act of war, and not in a position to do any direct damage to the United States.

These situations raise major questions of due process and executive authority. However, they make up only a tiny minority of drone attacks, which are mainly targeted at Al Qaeda and other militants in Pakistan, Afghanistan, Yemen, and elsewhere. The number of civilians in those countries killed by drones has also caused much debate. Yet Pakistan—which long professed official outrage about the drone attacks on its territory—subsequently made clear that it acquiesced in many of the attacks and, after long objecting to the civilian casualties, substantially lowered their death count.[23] In an effort to wind down the war on terror and reduce the risks of killing innocent bystanders, President Barack Obama proposed stricter guidelines on drone attacks and the transfer of responsibility for carrying them out from the Central Intelligence Agency to the Department of Defense. That plan, however, encountered unexpected opposition in Congress from those who think the CIA may actually do a better job of avoiding "collateral damage."[24]

Meanwhile, we are increasingly learning that drones are a fast-developing technology for policing purposes in many jurisdictions within the United States, prompting civil liberties concerns.[25] The deployment of drones is an excellent example of the blurring line between internal and external when it comes to contemporary uses of violence. Indeed, we may be witnessing the resuturing of police and military forces that had divided functionally in North Atlantic countries in the nineteenth century.

Notwithstanding the vigorous debate about the possible illegal uses of drones, it may be that we actually have more to fear from cyber-attack. Cyber-warfare involves many possibilities: hampering research and development, undermining the integrity of banking systems, knocking out air traffic control systems and thus threatening potentially thousands of passengers, disabling regional power grids, and more. The potential damage is nearly incalculable, yet a cyber-attack would bear little similarity to warfare understood as a face-to-face activity. Sabotage is old and familiar, of course, but until recently it has normally involved the physical presence of the attackers. That is clearly unnecessary in the case of cyber-warfare, even if the attackers also seek to remain anonymous and undetected. In the annual threat assessment presented by American intelligence officials to the Senate Intelligence Committee in early 2013, the director of national intelligence argued that a major cyber-attack is a greater immediate threat to the country than international terrorism—the first time any other threat has been ranked more highly than terrorism since 9/11.[26] Cyber-attacks are known to have been carried out by the Israelis and Americans against Iran (Stuxnet), while a new cold war has emerged between the United States and China as a result of cyber-warfare.[27] Recent attacks have indicated that the perpetrators may be at least as interested in destroying computer systems as they are in stealing data.[28] In early 2014, it was revealed that American cyber-warriors have implanted software in some hundred thousand computers worldwide that allows them to spy on those computers and conduct cyber-attacks.[29] Despite a broader climate of defense budget cutting, cyber-warfare is also one of the few areas where spending is likely to grow in the years to come.[30]

Finally, in the ultimate futuristic development, robots are assuming an ever-larger role in the conduct of military operations, locating improvised explosive devices (IEDs), assessing threats, and identifying targets. Commentators have begun to worry that we may soon have robots shooting targets without human intervention in the process at all.[31] The prospect of having robots fight our wars for us raises the threat of "riskless warfare"[32] even more sharply than does the use of drones.

With the implosion of the Soviet Union, the imbalance of power between the United States and the rest of the world became unprecedented, even though American defense spending as a percentage of gross domestic product fell dramatically during the Clinton years.[33] The terrorist attacks reversed those cuts, and the invasion of Iraq and Afghanistan led to unprecedented military expenditures, higher than anything ever seen before. As the wars in Iraq and Afghanistan have wound down, the American economy continues to face extraordinary pressures, especially in the aftermath of the early 2013 budget sequester and its automatic cuts to government outlays. With the Tea Party pushing the Republicans to the right, defense budget cuts—long regarded as anathema in conservative circles—began to be viewed as acceptable among Republicans.[34] The pressures to reduce the heft of the military in American life have become extraordinary.

In this context, the calls for a changing American way of warfare—including more use of warrior-free technology—became increasingly loud. The United States will now seek increasingly to achieve its military and foreign policy objectives "by, with, and through" partners and allies rather than by itself.[35] Under the Obama administration, this has also involved a shift (back) to greater support for international institutions such as the International Criminal Court. According to Harold Hongju Koh, legal adviser to the State Department, international criminal justice institutions "can help increase stability and thus decrease the need for more costly military interventions in the future."[36] Finally, the Pentagon has come to view social science research as increasingly central to its operations. A former director of the Defense Intelligence Agency, Lieutenant General Michael Flynn, recently wrote that, because "the perceptions of populations are increasingly the center of gravity of all conflicts," sociocultural research is crucial "to preventing the onset of conflict, to effectively prosecuting conflict if it comes, and to ensuring attainment of political goals and sustainable peace after the end of conflict."[37] One of the Army's "flashiest new concepts" is the deployment of "regionalized brigades," whose soldiers would receive cultural and language training appropriate to their mission.[38]

The shift to what one might call remote-control dominance by the United States can be illuminated by the approach to recent crises in Libya and Mali. This approach has been referred to as the Obama doctrine, but that is because he is simply the first to have been forced to operate under the conditions of austerity and foreign policy caution created by the misadventures of the George W. Bush presidency. In Libya, the Obama administration sought to enact its recently announced commitment to "atrocity prevention"[39] by intervening on the side of the Libyan rebels. Amid considerable uncertainty about who those rebels were and wariness about getting involved in another ground conflict in a predominantly Muslim country, the Obama administration promised only very limited engagement, with no American boots on the ground. Some dismissed the intervention as neocolonial, but the Arab League officially requested outside intervention, and anti-Gaddafi Libyans indicated considerable gratitude after the fact. The United States was best equipped to provide logistical and intelligence support to the rebels, but command and control remained in the hands of NATO, which has been in search of a mission since the end of the Cold War, when it was famously meant to keep the Americans in, the Russians out, and the Germans down. In general, the European Union continues to be unable to mount a serious "foreign and security policy," which many have hoped for a long time that they would do.[40] Obama sought to do something positive with American power but shrank from leading the charge.

The Mali crisis of 2013 further reveals the limits and reticence of American foreign policy in the post-Bush era. The Sahara has traditionally been the province of European imperial endeavors, not American ones, although the presence of oil in the region unavoidably drew in American activity. In 2012, however, the democratically elected president of Mali was toppled by a military coup involving, among others, soldiers trained by the United States as part of various American initiatives to forestall the development of terrorism in the area.[41] The northern part of the country was unsettled by conflict between Islamists and Touareg independence forces, seeking greater autonomy from the capital, Bamako. But when Islamist militants began imposing a strict version of sharia law in the north, Mali suddenly got on the map for reasons other than its fabled musicians (e.g., Ali Farka Touré and Salif Keita) and its Saharan version of the New Orleans Jazz and Heritage Festival.[42] The conflict greatly escalated when the French president responded to an urgent request from the head of Mali's government to intervene. The former colonial power in the country sent in several hundred troops, who were expected to stiffen the resolve of the unreliable Malian army, whose soldiers had a tendency to cut and run rather than stand and fight. Yet the French troops quickly found themselves in a firefight with Islamists. The United States had little interest in such a boots-on-the-ground intervention and had assumed it could rely on the Malians and other West African troops mobilized by the Economic Community of West African States. The Islamist insurgents then perpetrated an unexpected attack on an oil facility far to the northwest in Algeria, taking European and American hostages and drawing the Algerians into the fight. The Algerians have a long history of dealing brutally with Islamists, and they responded without coordination with the United States or Europe. Suddenly Washington's desire to maintain calm in the region by way of such mechanisms as the Trans-Sahara Counterterrorism Partnership looked unpersuasive. Whether prudently or not, France has been far more inclined to intervene directly in the region than has the United States. As in the case of Libya, the United States has been more inclined to rely on regional military partnerships and a network of bases it has established to conduct surveillance—which is in fact largely being carried out by private contractors.[43] All of this is far from the sort of aggressive imperialism one associates with European colonialism or with the United States' earlier interventions in, say, the Dominican Republic in the 1950s, Vietnam in the 1960s, or Iraq and Afghanistan in the 2000s.

Conclusion

The citizen-soldier is an increasingly scarce phenomenon in a post–Cold War world in which the use of massed military force has become less central to global dominance than it had been previously. Today's conflicts and threats in the rich world do not involve land wars requiring millions of conscripted soldiers; rather, they mainly involve threats to which the chief responses are technological and sociological, not military in the traditional sense. Violence has increasingly migrated to the world's poorer areas. Accordingly, warfare involving wealthy countries will increasingly require few personnel and much sophisticated technology. American dominance, to the extent that it persists, will be maintained more by remote-control attacks on armed combatants and supportive civilian institutions rather than by direct confrontation with enemy soldiers. In the meantime, while the country has signed on to a UN effort to tie the sale of arms to human rights considerations, it can scarcely regulate arms sales within its own borders to achieve reductions in gun-related deaths.[44] The ideal of the citizen-soldier, never as great an empirical reality in the United States as it was a norm in the republican mind, is likely to recede further as professional militaries, special operations forces, cyber-warriors, and robots increasingly do the work of war carried out by the world's wealthier countries.

Notes

1. George Mosse, *Fallen Soldiers: Reshaping the Memory of the World Wars* (New York: Oxford University Press, 1990).
2. David Parrott, *The Business of War: Military Enterprise and Military Revolution in Early Modern Europe* (New York: Cambridge University Press, 2012), 56 (emphasis added).

3. Michael Mann, *The Sources of Social Power*, vol. 2 (New York: Cambridge University Press, 1993), 60–61.

4. Sarah Percy, *Mercenaries: The History of a Norm in International Relations* (New York: Oxford University Press, 2007), 166.

5. UN General Assembly, "International Convention against the Recruitment, Use, Financing and Training of Mercenaries," December 1989, available at http://www.un.org/documents/ga/res/44/a44r034.htm.

6. See the Militia Act of 1792, Article I, available at http://www.constitution.org/mil/mil_act_1792.htm.

7. R. Claire Snyder, "The Citizen-Soldier Tradition and Gender Integration of the U.S. Military," *Armed Forces and Society* 29, no. 2 (2003): 187–188.

8. James J. Sheehan, *Where Have All the Soldiers Gone? The Transformation of Modern Europe* (Boston: Houghton Mifflin, 2008), 178.

9. Arno Mayer, *The Persistence of the Old Regime: Europe to the Great War* (New York: Pantheon, 1981).

10. Ernst Jünger, *Storm of Steel* (New York: Penguin, 2004), originally *In Stahlgewittern*, 1920.

11. Robert Graves, *Good-bye to All That* (1929; repr., New York: Anchor Books, 2004); Sigmund Freud, *Beyond the Pleasure Principle* (1920; repr., New York: Norton, 1961).

12. Michael Cohen, "A Reality-Based Army," *Democracy: A Journal of Ideas* 33 (Summer 2014), available at http://www.democracyjournal.org/33/a-reality-based-army.php.

13. See Melissa Eddy, "Austrians Appear to Reject Changes to Conscript Army," *New York Times*, January 20, 2013, available at http://www.nytimes.com/2013/01/21/world/europe/austrians-appear-to-reject-changes-to-conscript-army.html.

14. Sheehan, *Where Have All the Soldiers Gone?* 178.

15. David Segal, *Recruiting for Uncle Sam: Citizenship and Military Manpower Policy* (Lawrence: University Press of Kansas, 1989), 45.

16. Quoted in Ronald R. Krebs, "The Enduring Citizen-Soldier Tradition in the United States," in *The New Citizen Armies: Israel's Armed Forces in Comparative Perspective*, ed. Stuart A. Cohen (New York: Routledge, 2010), 8.

17. Krebs, "The Enduring Citizen-Soldier Tradition."

18. Krebs, *Fighting for Rights: Military Service and the Politics of Citizenship* (Ithaca, NY: Cornell University Press, 2006), 190.

19. Jason Ukman, "British Army to Shrink to Smallest Size in a Century," *Washington Post*, July 18, 2011, available at http://www.washingtonpost.com/blogs/checkpoint-washington/post/british-army-to-shrink-to-smallest-size-in-a-century/2011/07/18/gIQAuUsrLI_blog.html; Steven Erlanger, "Grim Economics Shape France's Military Spending," *New York Times*, April 29, 2013, available at http://www.nytimes.com/2013/04/30/world/europe/grim-economics-shape-frances-military-spending.html.

20. Dinah Walker, "Trends in U.S. Military Spending," Council on Foreign Relations, July 15, 2014, available at http://www.cfr.org/geoeconomics/trends-us-military-spending/p28855.

21. Deborah Avant and Renée de Nevers, "Military Contractors and the American Way of War," *Daedalus* 140, no. 3 (2011): 88.

22. Wendy Ruderman, "To Stem Juvenile Robberies, Police Trail Youths before the Crime," *New York Times*, March 3, 2013, available at http://www.nytimes.com/2013/03/04/nyregion/to-stem-juvenile-robberies-police-trail-youths-before-the-crime.html.

23. Declan Walsh, "In a Surprise, Pakistan Says Fewer Civilians Died by Drones," *New York Times*, October 30, 2013, available at http://www.nytimes.com/2013/10/31/world/asia/pakistan-drone-strikes.html.

24. Eric Schmitt, "Congress Restricts Drone Program Shift," *New York Times*, January 16, 2014, available at http://www.nytimes.com/2014/01/17/us/politics/congress-restricts-drones-program-shift.html. For a general overview of the positions concerning drones, see Peter Bergen and Daniel Rothenberg, eds., *Drone Wars: Transforming Conflict, Law, and Policy* (New York: Cambridge University Press, 2015).

25. Nick Paumgarten, "Here's Looking at You," *New Yorker*, May 14, 2012, available at http://www.newyorker.com/reporting/2012/05/14/120514fa_fact_paumgarten.

26. Mark Mazzetti and David E. Sanger, "Security Leader Says U.S. Would Retaliate against Cyberattacks," *New York Times*, March 12, 2013, available at http://www.nytimes.com/2013/03/13/us/intelligence-official-warns-congress-that-cyberattacks-pose-threat-to-us.html.

27. David E. Sanger, "In Cyberspace, New Cold War," *New York Times*, February 24, 2013, available at http://www.nytimes.com/2013/02/25/world/asia/us-confronts-cyber-cold-war-with-china.html; see also Jane Perlez, "U.S. and China Put Focus on Cybersecurity," *New York Times*, April 22, 2013, available at http://www.nytimes.com/2013/04/23/world/asia/united-states-and-china-hold-military-talks-with-cybersecurity-a-focus.html.

28. Nicole Perlroth and David E. Sanger, "Cyberattacks Seem Meant to Destroy, Not Just Disrupt," *New York Times*, March 28, 2013, available at http://www.nytimes.com/2013/03/29/technology/corporate-cyberattackers-possibly-state-backed-now-seek-to-destroy-data.html.

29. David Sanger and Thom Shanker, "N.S.A. Devises Radio Pathway into Computers," *New York Times*, January 14, 2014, available at http://www.nytimes.com/2014/01/15/us/nsa-effort-pries-open-computers-not-connected-to-internet.html.

30. Mike Hoffman, "Cyber Is Likely Winner of 2015 Budget," *Defense Tech*, February 24, 2014, available at http://defensetech.org/2014/02/24/cyber-is-likely-winner-of-2015-budget/.

31. Bill Keller, "Smart Drones," *New York Times*, March 16, 2013, available at http://www.nytimes.com/2013/03/17/opinion/sunday/keller-smart-drones.html.

32. P. W. Singer, "Robots at War: The New Battlefield," in *The Changing Character of War*, ed. Hew Strachan and Sibylle Scheipers (New York: Oxford University Press, 2011), 350.

33. Stockholm International Peace Research Institute Military Expenditure Database, available at http://milexdata.sipri.org/result.php4.

34. Jonathan Weisman and Ashley Parker, "Acceptance of Defense Cuts Signals Shift in G.O.P. Focus," *New York Times*, February 24, 2013, available at http://www.nytimes.com/2013/02/25/us/politics/democrats-and-republicans-miscalculate-on-automatic-cuts.html.

35. Thom Shanker, "Military Sees Broader Role for Special Operations Forces, in Peace and War," *New York Times*, April 2, 2013, available at http://www.nytimes.com/2013/04/03/world/military-plans-broader-role-for-special-operations.html.

36. Marlise Simons, "U.S. Grows More Helpful to International Criminal Court, a Body It Once Scorned," *New York Times*, April 2, 2013, available at http://www.nytimes.com/2013/04/03/world/europe/us-assists-international-criminal-court-but-still-has-no-intention-of-joining-it.html.

37. Michael Flynn, "Preface," in "Operational Relevance of Behavioral and Social Science to DoD Missions," ed. Sarah Canna (NSI white paper, March 2013), 2, available at http://www.fabbs.org/files/7613/6396/3846/U_Social%20Science%20White%20Paper%20Approved%20for%20Public%20Release%2014Mar13%20Final.pdf.

38. Gordon Lubold, "America's Emaciated Army," *Foreign Policy*, January 16, 2014, available at http://www.foreignpolicy.com/articles/2014/01/16/the_us_military_is_slated_to_shed_150000_soldiers_can_it_still_go_to_war_with_so.

39. Pursuant to Presidential Study Directive 10 of August 4, 2011, President Obama in April 2012 announced the creation of an official Atrocities Prevention Board, intended to prevent genocide and mass atrocities and to fulfill the country's commitment to the UN mandate concerning a "responsibility to protect" those in danger of such harms. White House, "Fact Sheet: A Comprehensive Strategy and New Tools to Prevent and Respond to Atrocities," April 23, 2012, available at http://www.whitehouse.gov/the-press-office/2012/04/23/fact-sheet-comprehensive-strategy-and-new-tools-prevent-and-respond-atro.

40. For one prominent plea, see Jürgen Habermas and Jacques Derrida, "February 15, or, What Binds Europeans Together: Plea for a Common Foreign Policy, Beginning in Core Europe," in *Old Europe, New Europe, Core Europe: Transatlantic Relations after the Iraq War*, ed. Daniel Levy, Max Pensky, and John Torpey (New York: Verso, 2005), 3–13.

41. Lydia Khalil, "What Mali Says about U.S. and International Strategy in the Region," *New York Times*, April 16, 2014, available at http://www.nytimes.com/roomfordebate/2013/01/15/is-mali-a-new-line-in-the-sand-against-terror/what-mali-says-about-us-and-international-strategy-in-the-region.

42. See "Festival au Désert," *Wikipedia*, July 21, 2015, available at http://en.wikipedia.org/wiki/Festival_au_D%C3%A9sert (accessed November 19, 2015).

43. Mark Mazzetti and Eric Schmitt, "U.S. Sees Hazy Threat from Mali Militants," *New York Times*, January 16, 2013, available at http://www.nytimes.com/2013/01/17/world/africa/us-sees-hazy-threat-from-mali-militants.html.

44. Neil MacFarquhar, "U.N. Treaty Is First Aimed at Regulating Global Arms Sales," *New York Times*, April 2, 2013, available at http://www.nytimes.com/2013/04/03/world/arms-trade-treaty-approved-at-un.html.

2 /

The Changing Character and Enduring Nature of War

The Collision of State and Substate Polities

Rob Johnson

War is waged with a political intent: to govern, control, coerce, punish, or protect. War derives from politics and is resolved through politics. Two essential elements of modern war are commonly misunderstood and their solutions misaligned. The first element is that, while the character of war has changed and is ever changing, often with new technologies and accompanying techniques, the nature of war—its fundamental essence—has not. Crucially, every war is characterized by the dynamic tension between the state and the people, but in civil wars this tension is more starkly evident. The second element is that attempts to stabilize faltering bureaucratic states with military force are inadequate when confronted with clan-based systems of governance supported by irregular armed groups. Failure to appreciate war's unchanging nature and that military force alone cannot stabilize regimes has caused significant challenges for Western powers in conflicts since 1945, although the problem has long historical antecedents.

The events of the first two decades of the twenty-first century have made clear that wars, and particularly insurgencies or insurrections, have lost none of their historic significance. While some scholars have argued that the post–Cold War world ushered in an era of "new wars"—stemming from deepening processes of globalization and the rise of identity politics—historians have demonstrated continuities with the past.[1] Armed nonstate actors—which some treated as if they were a new phenomenon indicating the demise or weakening of the state—exhibited many similarities to militant groups of the past in their struggles against sultanistic regimes of the Maghreb and Middle East and more recently against Western interventionist forces in Afghanistan and Iraq. Indeed, the collision between state actors and established clan and patrimonial authorities in the countries of the global South—which themselves were struggling to come to terms with the ingress of new ideas, technologies, economic conditions, and aspirations—reflects a long and historic contest over the configuration of political authority.[2]

Scholars of the Oxford Changing Character of War Programme in the United Kingdom have been consistent in asserting that war and the state continue to be important in our understanding of modern conflict. Despite the preference of many institutions and scholars for the semantic neutrality of "conflict" or "security studies" or the assertion that international norms have rendered war irrelevant or impossible, particularly since the end of the Cold War, it is clear that states and those movements that resist state powers are not only still central to recent conflicts but engaged in what is recognizably war. The state continues to be challenged, as recent insurgencies and insurrections demonstrate, and that only makes the study of the state and war, in its widest meaning, more important. An understanding of this context, the continually changing character of war, underpinned by its enduring and unchanging nature, is vital to any analysis of current and recently transformed warfare.

A brief chapter does not permit a full exposition of both continuity and change in the present condition of war, so the focus here is on just one issue: the clash between the state and nonstate authorities such as clans and tribes that has given rise to violent insurrections and corresponding counterinsurgency campaigns, set against the background of transformations in war wrought by new technologies and ideas. Given the costs and complexities in trying to stabilize bureaucratic states over a substructure of clan-based polities in conflicts in the first years of the twenty-first century, much consideration has been given by scholars, governments, international institutions, and professional military personnel to the means and ways of stabilization, beyond aid, development, and diplomacy. In Western thinking, air power, often manifest in the form of drone strikes, is being re-emphasized; policies of counterterrorism are replacing protracted counterinsurgency; and there is a new agenda of upstream engagement or defense diplomacy, in which faltering states can be assisted and supported to prevent their failure.[3] Western powers are concerned that failed states are havens of terrorism or generate warlordism or paramilitarism that, in turn, create humanitarian disasters and fuel civil conflict and even wider regional instability. The distribution of financial aid, sharing of intelligence assets, sale of arms and equipment, and deployment of training teams to support local military forces and their auxiliaries is the manifestation of these concerns in Western policy. The United Kingdom National Security Strategy of 2010, for example, described an "an age of uncertainty," reflecting an evident fear of revisionist aggression by rival states and widespread international terrorism by a new generation of eschatologically minded self-sacrificial warriors, eager to inflict mass casualties and, potentially, armed with weapons of mass destruction.[4] The U.S. National Security Strategy also recognized an era of sweeping change and the pressing need to continue to combat international terrorism around the globe.[5]

Too often, national defense policies are shaped by domestic agendas, selected and sometimes exaggerated fears, and the aspirations of governments to fulfill a role on the world stage—all of which take insufficient notice of local drivers of conflict. During the Cold War, many regional conflicts were interpreted through the lens of ideology, a prism that tended to obscure any other factors. It is important to stress the importance of understanding the position and view of indigenous populations as well as the organic and historic nature of the "human terrain," which might consist of years of antagonism against exploitative state structures, and the situation of those that states tend to designate and essentialize as the "enemy."[6] This is not just the product of the cultural turn in scholarly studies but an observation of how states function, especially in civil wars; how states' military operations proceed; and how they are opposed. States are not just institutions that simply impose themselves with their own policies intact; they are mirrored by nonstate groups, which also have policies and practices, and both sides are affected by the other. Just as states at war use labor, resources, work, and ideas to mobilize, assert, and legitimize, so too do nonstate actors use their own channels of funding, systems of patronage, and tools for popular mobilization. Better understanding as to the interaction of these forces gives us a greater insight into what generates the familiar and historic dynamic of war.

The enduring nature of war has been well documented by scholars and practitioners. Europe's experience of the two totalizing and global wars was catastrophic in lives and economy. Precisely because of war's grim nature, during those wars, new regulation and constraints were applied to it: until 1945, wars had to be declared, certain weapons were and still are banned, and conventions protected prisoners, the wounded, and noncombatants. Since then, the United Nations Charter and various conventions and protocols have been established to protect and preserve noncombatants. But success has been limited in this regard. Thousands of civilians have died in conflicts since 1945 and far more, proportionally, than regular military forces.[7] The experience of war has remained as brutal and callous as ever. As Chris Coker so wisely notes, the delusion that war *is* more constrained and limited today makes the temptation, and thus the ability, to wage it far easier.[8] In every conflict, the thresholds against atrocities, so quaintly delineated by wise counsels in peacetime, are crossed, and savagery escalates as war progresses, as Carl von Clausewitz observed in *On War*, toward its theoretical absolute form and function.[9]

Revolutionary Warfare

In the *Transformation of War*, Martin van Creveld argues that the form of governance and adherence to accepted laws could determine the degree of violence in war, and thus states might opt for self-restraint.[10] But when states engaged a party with no government, there was little distinction between war and peace and thus a protracted, attritional, parasitic conflict would ensue. While van Creveld is broadly accurate, in many conflicts states are just as capable of waging protracted, attritional, and unrestrained war as nonstate actors. Indeed, the idea that democratic states wage war in a more restrained way is a fallacy.

In civil wars, the thresholds of restraint and the boundaries that differentiate and protect civilians are crossed more readily. Typically states faced by elusive guerrilla enemies begin to escalate their use of force and assert their power through new draconian laws that, in turn, alienate the population whom they govern and for whom they are responsible. The guerrillas seek to provoke the state into heavy-handedness, to discredit the government by demonstrating its inability to govern or protect, but they will also intimidate and discipline the population to maximize their security and allegiances.

In the 1990s, Mary Kaldor posited that a "new type of organized violence developed, especially in Africa and Eastern Europe":

> New wars involve a blurring of the distinctions between war (usually defined as violence between states or organized political groups for political motives), organized crime (violence undertaken by privately organized groups for private purposes, usually financial gain) and large-scale violations of human rights (violence undertaken by states or political groups against individuals).[11]

At the end of the Cold War, it was understandable that conflicts occurring outside the old zones of confrontation might be seen as new, particularly when the bipolar constraints that had contained violence were released. In the same period there was awe at new technologies and speculation about their implications. The revolution in military affairs, now called transformation, promised the elimination of friction in war.[12] It was asserted that information technology and satellites would allow Western forces to see their adversaries and communicate better, move faster, make quicker decisions, bring in overwhelming and precise firepower, and discriminate between civilians and combatants. Western forces would be able to break down enemy organizations, particularly their command and control systems, while eradicating the error of human emotion and stress on their own side. This was to be pure war, without politics.

Thomas Hammes, in contrast, believes that it is, in fact, old methods that will really matter. Despite identifying a new "fourth-generation warfare," the notions he advances are quite familiar:

> Fourth-generation warfare uses all available networks—political, economic, social, and military—to convince the enemy's political decision makers that their strategic goals are either unachievable or too costly for the perceived benefit. Unlike previous generations of warfare, it does not attempt to win by defeating the enemy's military forces. Instead, it attacks the political will and morale of the enemy.[13]

Nevertheless, the ways and means of fighting a guerrilla conflict have been far less important than the political ends. The insurgents or guerrillas must try to exhaust the strategic patience of their state adversaries, especially democratic ones, by imposing unacceptable costs in lives, treasure, and reputation if they are to succeed. They must also exploit any alternative governance structures, either reverting to old clan, sectarian, or particularist allegiances or establishing entirely new transnational, transtribal hierarchies. Al Shabab in Somalia, the Taliban in Afghanistan, Al Qaeda in Iraq, and the ISIS (Islamic State of Iraq and Syria) movement are examples of the latter in the 2010s.

Recent conflicts indicate that technological superiority does not guarantee the defeat of an enemy's will to resist. The destruction of command and control systems has created protraction by default because it drives the dispersal of an enemy's forces into the security of the civilian population. Defeating conventional forces or killing significant numbers of irregulars has not guaranteed an adversary's emotional collapse. Technological victory cannot eradicate an identity even when it can cause a temporary loss of will to fight. The utility of force is in creating a neutralizing effect on one's adversary rather than merely inflicting casualties, although that too can still be important. Given the sheer lethality and effectiveness of modern weapon systems, dispersal and devolved command is essential and has been particularly acute whenever one force is disproportionately overmatched by another. Guerrilla warfare, recognized by Clausewitz as something to be avoided by conventional forces by means of decisive action, nevertheless reappears throughout history with some frequency, and its drivers, methods, and countermeasures are well rehearsed.[14] Guerrillas must target routes and communications, which they recognize as the main vulnerability of armies. They have to "hug" conventional forces in combat to prevent them from being able to use their superior weapons and then quickly disperse again into the civilian population to survive. They must fight using the emotional effect of casualties, wearing down their opponent's will. "Asymmetrical war," an imprecise term that nevertheless illustrates that there is no equilibrium in war, is merely the extension of modern war. Its main characteristics are dispersion, stealth, and devolution, but its purpose, as always, remains political in nature.

The current character of war for advanced states, we observe, is of increasing computerization, with an emphasis on the metrics of targeting, firing, surveillance, and neutralizing effects. Insurgents endeavor to overload command and communication systems by multiple firing points or various forms of attack, including suicide bombers. Special Forces teams are, of course, still required to carry out close surveillance to enable the computerized weapons to engage, and they often need to be concealed inside populations or to recruit local auxiliaries by engaging men using a high degree of empathy and understanding of the needs of nonstate actors and their agendas.[15] Despite attempts to eliminate friction, human personnel and their high-tech systems are still vulnerable when countering terrorism and insurgency to exhaustion, technical failure, and the erroneous decisions taken by tired, stressed, and scrutinized commanders. Information fog may be less of an obstacle in conventional warfare, but insurgents try to confuse, obscure, and remain concealed. The high tempo of conventional war suits the high-tech systems of Western forces, but periods of protracted warfare do not, because it is here that friction reasserts itself more strongly.

Critics of new-war proponents and the cult phrases associated with modern warfare have argued that too much emphasis is placed on what is apparently novel. Hew Strachan and Sibylle Scheipers demonstrate that apparently new actors in war do not alone constitute change in war; war is more multidimensional, enduring, and complex than that.[16] The motivations and agendas of nonstate belligerents are familiar: Afghan irregulars of the eighteenth century, the *eljaris*, went to war against infidels to fulfill a sacred cause but sought to avoid battles in order to concentrate on looting, and they concealed themselves among civilian populations the moment they came under a significant threat. Material opportunism and ideologically inspired Afghan irregulars, which characterized the conflict in Afghanistan in 2001–2014, are not new.

The essence of both insurgency and war is unchanging, but aspects of the character of modern warfare, aside from changes in technology and language, would be familiar to historical actors too, including the need for the strategic correlation of ends, ways, and means; the enduring importance of command and leadership; the need to maintain morale during protracted operations; the difficulties of sustaining resistance amid heavy losses; the thorny political problems of suppressing internal revolts led by (armed) civilians; and the common asymmetry of civil war. The experiences in the Balkans, the Caucasus, Iraq, Afghanistan, and the Arab Spring of the Maghreb and the Middle East since the 1990s indicate that the fundamental issues of insurrection, insurgency, and revolutionary war are not new and exhibit many patterns that can be identified from the past. The problem is that observers can be too selective, reaching for common tropes of people's war, state repression, or proxy warfare. What is needed is a more critical and rigorous analysis of insurgency and its relationship with substate or transnational politics if we are to assess what is genuinely transforming in war.

In the West, there is a predilection to invent paradigmatic terms that capture the essence of its approach to war and that hold out the promise of a swift, decisive victory against any adversary. There were claims in the 1990s that "full-spectrum dominance" would give the United States overwhelming military advantages in every sphere of land, sea, air, submarine, and electronic warfare. Confidence in the combination of destructive firepower and the ability to break the will to resist through the destruction of command and control systems gave rise to the epithet "shock and awe" in 2003. The intention to create a comprehensive "system-of-systems," announced by Admiral William Owens, the vice-chairman of the Joint Chiefs of Staff, was propelled by faith in a much-documented, American-led "revolution in military affairs," which pointed toward superiority in cyber-warfare as well as all the other domains of conflict.[17] Analysts have become so familiar with many of these concepts in their vocabulary that it is easy to accept them uncritically, and indeed, they become part of an oft-repeated taxonomy, reassuring those who use them that they somehow know war. After the war in Iraq, there was a corresponding preference for "stand-off" attacks (that is, the delivery of fire on to targets from the air or from a position that prevents exposure to the risk of retaliation or fighting at close quarters) on America's enemies, an aversion to risk or casualties, and increasing opposition to the idea of war among democratic voters. In other words, despite the sophistication of new technologies and systems, there was a growing gap between military capability and the reality of war for many civilian populations. Greater precision, for example, could not be separated from the full context of war.

There is a risk of a dangerous self-delusion of misplaced confidence in technology as the solution to war itself. In *Virtuous War*, James Der Derian argues there is a Western obsession with war technology but an increasing disconnection with the reality of war.[18] Dwight Eisenhower coined the phrase "the military-industrial complex," which Der Derian takes further to identify "the military-industrial-media-entertainment network," suggesting that the West is beginning to believe its own aspirations and ideals to the exclusion of all others. Full-spectrum dominance and advanced helicopters did not prevent the Black Hawk incident and the ensuing close-quarter street combat in Mogadishu in the 1990s. High-tech systems, especially in surveillance, were temporarily neutralized or hampered by bad weather in Iraq in 2003 and subsequently in the mountains of Afghanistan. In the future, there is no guarantee that cyber-war will be entirely bloodless or function as smoothly as its champions claim.

The desire to seek out technological and scientific solutions to war fighting is understandable, but it cannot come at the expense of the study of conflict between states and their peoples. Fresh thinking is required. Too many have accepted uncritically the assurances of revolutionary writers like Karl Marx, Vladimir Lenin, Mao Tse-tung, Che Guevara, and Carlos Marighella about the conduct of revolutionary war, and there is a need to resituate the factors of guerrilla war, insurgency, and substate conflict, such as the relationships of time and space, coercion and popular consent, will and intent, reputation and charisma, resources and sustainability, and substate governance and power.

The Responses

Some have suggested that the advanced nature of the West drives rivals to seek military advantage by other means. Shawn Brimley of the U.S. Army War College warns, "America's continued strength in major force-on-force conflict will incentivize future adversaries toward distributed cellular forms of insurgency characterized by the improvised explosive devices and ambushes, seen in Iraq and Afghanistan, or toward the use of organized small-unit kinetic operations buttressed with employment of advanced technology such as the antitank and antiship munitions successfully employed by Hezbollah in the 2006 Lebanon War."[19]

Officially there has been recognition of rivals' and potential enemies' need to find weaknesses in the West. The U.S. National Defense Panel in 1997 noted, "We can assume that our enemies and future adversaries have learned from the Gulf War. They are unlikely to confront us conventionally with mass armor formations, air superiority forces, and deep-water naval fleets of their own, all areas of overwhelming U.S. strength today. Instead, they may find new ways to attack our interests, our forces, and our citizens. They will look for ways to match their strength against our weaknesses."[20] There was even some assessment of the changed strategic environment. The U.S. National Intelligence Council's *Global Trends 2025* report suggests, "The international system—as constructed following the Second World War—will be almost unrecognizable by 2025 owing to the rise of emerging powers, a globalizing economy, an historic transfer of relative wealth and economic power from West to East, and the growing influence of nonstate actors."[21] Apocalyptic assessments of the future, dominated by terrorism, contamination, pandemic, famine, and flood have been accompanied by bleak but assertive predictions about the demise of not only the international system but also the nation-state.

Confident predictions about the demise of the state or of war as a policy option nevertheless seem premature. History suggests that states will fail and be replaced, often amid much conflict. Chris Cramer argues that insurgencies and civil wars serve a purpose of state renewal and can, if they run their course without intervention, produce a "better peace." Ricardo Soares de Oliveira and James Mayall note that illiberal democracy, after such wars, can also be a better guarantor for preserving peace.[22]

Leading Western states have created an international system that serves their interests. Those that fail to benefit from this system seek to change it, expressing anger and frustration, protesting, arming, and adapting themselves against foreign occupation, internal state failure, or gross official corruption. The West is quick to condemn the emergence of a new nexus of organized crime, terrorism, and insurgency and has sustained its appetite for military interventions, sometimes under the guise of humanitarian intervention, as peace enforcement missions for the UN, and more explicitly in the pursuit of their national interests.[23] Jeremy Black has argued that war must be resituated in its sociocultural dimension, because it is not waged purely for national or state interests even though it is still intensely political in nature.[24] Indeed, noting the importance of civil war, he has pointed out that regular armed forces were invariably raised to control populations, thus maintaining the state and its economic system. The predictability and security of the United States' economy and its ability to contain the inherently destabilizing pressures of democracy mean that it could wage war confidently beyond its own borders. Black suggests that, in the future, increasing resource competition, urbanization, and population growth will jeopardize the stability of many states and confront the West with many more scenarios like Afghanistan, Iraq, Somalia, or Syria.[25] There is likely to be a proliferation of nonstate belligerents, who, as in such movements in the past, are unlikely to accept the norms of war and measures of restraint imposed by the dominating powers. As Steven Metz notes, "The best legal system on earth matters little if it is consistently ignored."[26]

The inability to confidently predict the future explains the confusion about the posture of Western military forces, especially the hegemonic United States. In "Network Centric Warfare," a report issued in 1998, the U.S. Department of Defense's transformation guidance for the future ignored the historical processes or contexts of war and argued that enemies in the twenty-first century would ignore the past and willingly fight the United States in a high-tech, fast-paced-maneuver campaign that reinforced American strengths while avoiding its weaknesses. Anxiety about protracted war, so prominent after the Iraq and Afghan insurgencies, nevertheless elicited volumes of work that reached back to the methods of counterinsurgency used by the European powers during wars of decolonization. Thus, while the 1998 report had defined asymmetry simply as "not fighting fair,"[27] by 2006 analysts and military practitioners were reaching for old doctrines that could provide solutions to such ancient concepts as ambush along the line of march, assassination, sniping, raiding, and insurrection. These were then republished and presented as "new," with reassuring paragraphs about the ability to conduct achievable operations during "wars amongst the people."[28]

The logic of a proposed "light footprint" in Western expeditionary warfare was to remain agile, minimize the burden of logistics, and avoid the antagonism of the local people toward any overt and large-scale military presence.[29] In the early years of its mission in Afghanistan, in 2001–2003, the United States sought specifically to avoid any idea of occupation in Afghanistan in order to prevent repetition of the Soviet mistake in 1979. In 2001 there was considerable faith in the ability of air power to deliver solutions without a substantial ground commitment.[30] In fact, the logic of smaller ground forces is greater vulnerability and minimal intelligence that can be compensated only by a greater reliance on air power. Yet despite the advent of precision strike and enhanced targeting, reliance on air power has been the cause of higher civilian casualties. This proved counterproductive in the militarized policing operations Western forces found themselves in.

Critically, an expeditionary posture and air power could not provide security for the establishment of a new government. After operations against Libya in 2011, there was again enthusiasm for air operations that could avoid a ground commitment, even though air or missile attacks against state and failed-state systems in the 1990s against Iraq and Afghanistan did not, on their own, achieve lasting results. Tragically, the attempts at air policing from as early as the 1920s made it apparent that this approach was limited.[31] It has taken some time for the Western powers to realize that not only their methods of war fighting but also their campaign designs, including doctrines, could not be treated as immutably superior, and they have been forced to change constantly as Iraq and Afghanistan operations unfolded. Moreover, the absence of a political solution to the occupations and local grievances gave rise to insurgency and guerrilla war and exposed the inadequacy of Western approaches to revolutionary war and state stabilization.

New technologies, such as unmanned aerial vehicles (UAVs, or drones) and robotics, and new methods, such as denial of service or disruption attacks using cyber technology, do no more to advance the idea of stabilization or governance than did the faith in air power in the early twentieth century and appear to have little relevance in ending insurgencies in developing countries. Debate has raged on the character, legal and ethical or not, of targeted killing within states that are not at war with the West, such as Yemen, Syria, or Pakistan; of temporarily removing insurgent fighters from the battlefield by extralegal incarceration; and of extraordinary rendition and interrogation techniques used on suspected fighters.[32] The Western concern to protect populations and the rights of the individual, deeply internalized from the advent of massed air bombardment in World War I and the annihilationist strategies of the total wars, is not the priority for many non-Western belligerents. Disturbing and unpalatable though it may be for the West, the fact is that intimidation, fear of reprisals, and overwhelming military power have sometimes swayed a population against the insurgents rather than the limited war and ethical behavior so treasured by Westerners. Conversely, the Western policing approach with drone strikes erodes the boundaries between war and peace still further and makes it easier for nonstate groups to assert that they, too, possess the right to strike back in an international setting.

Engaging in wars across national borders and attempting to stabilize polities is redolent of the debate about "turbulent frontiers" in the history of empires. J. S. Galbraith notes that when stable and dominant polities like empires encountered weak, clan-based systems, the imperial forces were drawn into wars they did not desire.[33] European imperialists, quick to condemn African and Middle Eastern lawlessness and despotism for the frequent collapse of order in their empires, often failed to realize that it was the very presence of the Europeans that caused destabilization and disorder.[34] They also failed to realize that local, nonstate actors had their own agendas, sometimes to exploit, resist, or subvert the presence of Westerners.[35] Their overriding concern was to engage in internal contests for power, asserting themselves in alignment with Western interests or in defiance of them. Similar characteristics can be found in the conflicts in Iraq and Afghanistan.[36] In interventions, collaboration is common, and so too is hedging by local populations, outbreaks of resistance, and ruthless opportunism to acquire resources. Equally, attempts by state powers to control proxy irregular armed groups can unravel quickly in the face of local agency.

Insurgents have been historically the most successful when they were able to provide a viable alternative governance and eventually replace the existing apparatus. However, in many cases in the early twenty-first century, the rule of insurgents in territories they control is characterized as warlordism, exploitative criminal activity, or coerciveness through ideological or sectarian intolerance. In the past, successful counterrevolutionary or counterinsurgent action at the state level depended on three broad options: swift and early armed intervention that prevented communalism from developing into a protracted conflict, the immediate shift to a political process to remove the seat of the grievance, or a concessionary approach to power sharing to prevent violence developing in the first place. All three approaches were used by the British during their gradual withdrawal from imperial rule. At the substate level, successful states identified local hierarchies and endorsed groups or individuals, investing them with authority, posts in the executive, and resources. While not foolproof, engaging peripheral and minority groups and encouraging reconciliation through a new national identity was invariably positive.

As a means of last resort, Western states will continue to wage war in the future to protect populations and to preserve international institutions, transnational ideas of liberty, and the functioning of a global economy that brings the greatest good to the largest number. The character of war will be transformed, but it may not always be decided by new technological developments. Trends suggest that the privatization of violence will be important, which in turn suggests that nonstate armed groups will be of even greater significance but equally that democratic governments will be under immense pressure to protect the lives of individuals threatened or held captive overseas.[37] War will be conducted with greater precision, but correspondingly belligerents will need to be more stealthy, concealed, located in depth, and dispersed. There will be greater emphasis on identifying the nodal-systemic vulnerabilities of the adversary, but traditional matters of morale, will, and intent are sure to remain vitally important.

The grim realities of war are unlikely to change, even though their forms will evolve. Some will be familiar. Violent transnational activism to mobilize populations against Western principles, intense actions at close quarters to avoid air power, and suicide terrorism seem very likely. The new forms might include the greater use of social media for mobilization, deniable proxy detonations of weapons of mass destruction, or suicide attacks by troops already condemned by the AIDS virus and with literally nothing to lose. Warfare is likely to be individualized further as smaller and smaller groups assert the right to wage war, equipped with significant combat power that was never available to our historical ancestors. Global paramilitary-style policing and the deployment of nomadic robotic weapons engaging fighters or unarmed civilians implicated in terrorism or insurgency have been imagined by Manuel de Landa.[38] Recognized historical aspects of war, with accepted status and rituals and outward forms of identification and restraints, may not survive.

There is little that is new in the nature of war: its essence is largely unchanging, in the past, the present, or, as far as we can tell, the future. Thucydides would have recognized the suffering of the civil war in Syria after 2011. Clausewitz identified war in its essential state as unrestrained and dynamic and the trinity of its elements—passion, reason, and chance—as in a state of tension, not equilibrium. It was precisely because of war's tendency to break out of the rational confines of policy makers that he argued for the focus of all effort to force a decision. This could create conditions in the fighting in which war was not merely an extension of policy but often became policy.[39] Thomas Hobbes, too, believed that violence was elemental and could be constrained only by powerful institutions. Even Jean-Jacques Rousseau, who argued that war was not a natural state for humans, noted that men had to be herded into organizations of war by belligerent states, and these, of course, were both bureaucratic versions of a state and, as they are today, rather more akin to protostates and political movements.

There is a worrying tendency to impose principles to conflicts retrospectively, to sanitize and explain them through terms acceptable in the West, and thus to conceal the realities of war. The UN denies "war" in favor of "armed conflict"; there is talk of "humanitarian intervention" and "responsibility to protect" and regret at the unfortunate consequence of "collateral damage." Against these reassuring ideas, history makes uncomfortable reading. Western governments have periodically supported revolutionary forces to serve their own interests, only to find they had inadvertently fueled a civil war that resulted in civilian deaths. We should be aware, then, that consistent factors in causation and the conduct of revolutionary or counterrevolutionary war are overlooked to suit contemporary agendas. Tropes about revolutionary war, inherited less critically than they should have been, have created untested assumptions on the strengths and vulnerabilities of insurgent movements. Focusing on the *fighting* in guerrilla war has obscured the emphasis that should be placed on their attempts to build substitute political systems. Attempts to tackle insurgencies and revolutionary guerrilla wars with military means, in either a limited or escalatory fashion, have been less successful than more immediate political solutions that can fulfill some aspirations or, at the very least, take the heat out of a conflict. To tackle an insurrection among the urban poor in one of the vast megacities of South America, Africa, or Asia, for example, would require far more than a military operation and would involve the establishment of a mechanism for political dialogue, representation, policing, cooperative decision making over resource distribution, employment, and security, to name just a few of the most pressing tasks.

The failure of some states can be attributed to internal tensions and collapse, many of which are the result of an inflexibility or inability to absorb the effects of a system weighted in favor of dominant states. The clan-based, communal, or sectarian systems of governance that resist, subvert, or manipulate the imposition of bureaucratic models create opportunities to seize resources, exploit vulnerabilities, and avoid international responsibilities. Older systems are reasserted, and sometimes entirely new revolutionary movements emerge as a reaction to the impositions of state demands, injunctions, and security forces. The failure to stabilize a central government in Afghanistan, Somalia, or similar polities reflects the resilience of traditional forms of substate governance and the desire to avoid the consequences of a system that might impose restrictions and penalties or deny the fruits of local opportunism. They are a reminder that wars among the people, civil wars, and insurgency are not new but historic, and they reflect a traditional power struggle in which the rules of conventional war and the state system behind them are always undermined and challenged. They are also recognizably similar to wars in the past; in nature, they feature friction, violence, death, psychological stresses, and adaptation, and in character, they constantly evolve as the result of dynamic interactions. Politics is always the essence of war, and it is always the path to its resolution.

Notes

1. Mary Kaldor, *New and Old Wars: Organised Violence in a Global Era* (Stanford, CA: Stanford University Press, 1999); Stathis Kalyvas, "'New' and 'Old' Civil Wars: A Valid Distinction?" *World Politics* 54, no. 1 (2001): 99–100.

2. See Paul Collier and Anke Hoeffler, "Greed and Grievance in Civil War," World Bank Policy Research Paper 2355, World Bank, Washington, DC, 2000; Christopher Clapham, ed., *African Guerrillas* (Oxford: Oxford University Press, 1998); Anthony Clayton, *Frontiersmen: Warfare in Africa since 1950* (London: University College London Press, 1999); George Kieh and Rousseau Mukenege, eds., *Zones of Conflict in Africa: Theories and Cases* (Westport, CT: Praeger, 2002); Vo Nguyen Giap, *People's War, People's Army: The Viet Cong Insurrection Manual for Underdeveloped Countries* (Honolulu, HI: University Press of the Pacific, 2001); Olivier Roy, *Islam and Resistance in Afghanistan* (Cambridge: Cambridge Middle East Library, Cambridge University Press, 1986), 224; and Kamal Matinuddin, *The Taliban Phenomenon: Afghanistan, 1994–1997* (Oxford: Oxford University Press, 2000), 71.

3. State failure is a contested idea in academic circles: while there is widespread recognition of social, economic, or political indicators of an executive's ability to govern, William Easterly and Laura Freschi argue that "state failure" has no coherent definition and that various definitions are used to justify Western intervention. William Easterly and Laura Freschi, "Top 5 Reasons Why 'Failed State' Is a Failed Concept," *AidWatch*, January 13, 2010, http://aidwatchers.com/2010/01/top-5-reasons-why-%E2%80%9Cfailed-state%E2%80%9D-is-a-failed-concept.

4. Her Majesty's Government, "A Strong Britain in an Age of Uncertainty: The National Security Strategy," October 2010, available at https://www.gov.uk/government/uploads/system/uploads/attachment_data/file/61936/national-security-strategy.pdf.

5. White House, "National Security Strategy of the United States," May 2010, available at https://www.whitehouse.gov/sites/default/files/rss_viewer/national_security_strategy.pdf.

6. Anatol Lieven, *Pakistan: A Hard Country* (London: Allen Lane, 2011), 19–21.

7. James D. Fearon and David D. Laitin, "Ethnicity, Insurgency and Civil War," *American Political Science Review* 97, no. 1 (2003): 75–90.

8. Christopher Coker, interview by the author, Stockholm, March 2012.

9. Carl von Clausewitz, *On War*, trans. Michael Howard and Peter Paret (Princeton: Princeton University Press, 1976), chap. 1.

10. Martin van Creveld, *The Transformation of War* (New York: Free Press, 1991), 33; see also Martin van Creveld, *The Rise and Decline of the State* (Cambridge: Cambridge University Press, 1999), chap. 3.

11. Mary Kaldor, *New and Old Wars: Organised Violence in a Global Era*, 3rd ed. (Cambridge, UK: Polity Press, 2012), 1.

12. E. C. Sloan, *The Revolution in Military Affairs* (Montreal: McGill University Press, 2002).

13. Thomas X. Hammes, *The Sling and the Stone* (St. Paul, MN: Zenith Press, 2004), 190. See also Thomas X. Hammes, "The Evolution of War: The Fourth Generation," *Marine Corps Gazette*, September 1994, pp. 35–44.

14. Max Boot, *Invisible Armies* (New York: Norton, 2013).

15. Diane E. Davis and Anthony W. Pereira, ed., *Irregular Armed Forces and Their Role in Politics and State Formation* (Cambridge: Cambridge University Press, 2003), 149–177; Austin Long, "Going Old School: U.S. Army Special Forces Return to the Villages," *Foreign Policy*, July 21, 2010, available at http://afpak.foreignpolicy.com/posts/2010/07/21/going_old_school_us_army_special_forces_return_to_the_villages; Charles Tilly, *The Politics of Collective Violence* (Cambridge: Cambridge University Press, 2003), 19.

16. Hew Strachan and Sibylle Scheipers, eds., *The Changing Character of War* (Oxford: Oxford University Press, 2011).

17. W. H. Manthorpe Jr., "The Emerging Joint System of Systems: A Systems Engineering Challenge and Opportunity for APL," *Johns Hopkins APL Technical Digest* 17, no. 3 (1996): 310; William A. Owens, "The Emerging U.S. System-of-Systems," *Strategic Forum*, no. 63 (February 1996): 1.

18. James Der Derian, *Virtuous War* (Boulder, CO: Westview, 2001).

19. Shawn Brimley, "Crafting Strategy in an Age of Transition," *Parameters* 38, no. 4 (2009): 27.

20. National Defense Panel, "Transforming Defense: National Security in the 21st Century," December 1997, p. 11, available at http://www.dod.gov/pubs/foi/Reading_Room/Other/902.pdf.

21. National Intelligence Council, *Global Trends 2025: A Transformed World* (Washington, DC: U.S. Government Printing Office, 2008), vi. See also U.S. Joint Forces Command, "The Joint Operating Environment 2008: Challenges and Implications for the Future Joint Force," November 25, 2008, pp. 10–23, available at http://fas.org/man/eprint/joe2008.pdf; Robert Gates, U.S. secretary of defense, speech at the Economic Club, Chicago, July 2009.

22. Christopher Cramer, *Civil War Is Not a Stupid Thing: Accounting for Violence in Developing Countries* (London: Hurst, 2006); Ricardo Soares de Oliveira and James Mayall, *The New Protectorates: International Tutelage and the Making of Liberal States* (London: Hurst, 2011).

23. See David Welch and Jo Fox, eds., *Justifying War* (Basingstoke, UK: Palgrave Macmillan, 2012).

24. Jeremy Black, cited in Steven Metz and Phillip Cuccia, "Defining War for the 21st Century," 2010 Strategic Studies Institute Annual Strategy Conference Report, February 2011, p. 15, available at http://www.strategicstudiesinstitute.army.mil/pdffiles/PUB1036.pdf.

25. See Jeremy Black, *War and the New Disorder in the Twenty-First Century*, 2nd ed. (London: Bloomsbury Academic, 2004).

26. Metz and Cuccia, "Defining War for the 21st Century," 23.

27. Steve Metz and Douglas V. Johnson II, *Asymmetry and U.S. Military Strategy: Definition, Background, and Strategic Concepts* (Carlisle, PA: Strategic Studies Institute, 2001), 5.

28. See, e.g., Rupert Smith, *The Utility of Force: The Art of War in the Modern World* (New York: Knopf, 2007).

29. The origin of the phrase is contested but appears with some frequency after 1996. See, for example, Astri Suhrke, "The Case for a Light Footprint: The International Project in Afghanistan," 2010, available at https://www.soas.ac.uk/cccac/events/anthonyhyman/file58420.pdf.

30. D. M. Drew, "U.S. Airpower Theory and the Insurgent Challenge: A Short Journey to Confusion," *Journal of Military History* 62 (1998): 809–832.

31. David E. Omissi, *Air Power and Colonial Control: The Royal Air Force, 1919–1939* (Manchester, UK: Manchester University Press, 1990).

32. Kenneth Roth, "What Rules Should Govern US Drone Attacks?" *New York Review* 25 (March 2013): 16–18.

33. J. S. Galbraith, "The 'Turbulent Frontier' as a Factor in British Expansion," *Comparative Studies in Society and History* 11 (1959–1960): 151–168.

34. Ronald Robinson, "Non-European Foundations of European Imperialism: Sketch for a Theory of Collaboration," in *Studies in the Theory of Imperialism*, ed. Roger Owen and Bob Sutcliffe (London: Prentice Hall, 1972), 132.

35. Ranajit Guha, *Elementary Aspects of Peasant Insurgency in Colonial India* (New Delhi: Oxford University Press, 1983).

36. T. Dodge and S. Simon, eds., *Iraq at the Crossroads: State and Society in the Shadow of Regime Change* (Oxford: Oxford University Press, 2003); G. Robinson, "The Battle for Iraq: Islamic Insurgencies in Comparative Perspective," *Third World Quarterly* 28, no. 2 (2007): 267–273.

37. E. Krahmann, *States, Citizens and the Privatisation of Security* (New York: Cambridge University Press, 2010); T. Jäger and G. Kümmel, eds., *Private Military and Security Companies: Chances, Problems, Pitfalls and Prospects* (Wiesbaden, Germany: VS Verlag, 2007).

38. Manuel de Landa, *War in an Age of Intelligent Machines* (New York: Zone Books, 1991).

39. Hew Strachan, interview by the author, Oxford, March 2013.

3 /

Plus Ça Change

War and State Building

Ian Roxborough

Theories of "new wars" are right in asserting that, as conventional warfare between advanced states recedes into the background, the predominant form of armed conflict will be counterinsurgency and state-building operations. Theories of new wars are wrong in asserting that such wars are "new." This chapter uses the cases of U.S. intervention in the Philippines (from 1898 to the mid-1950s) and in Vietnam (during the late 1950s and early 1960s) to examine the dynamics of military-sponsored social engineering and state building. These episodes prefigured current efforts in Iraq and Afghanistan. The final part of the chapter discusses the general dynamics of state building and military social engineering and the implications for future warfare.

State building is still on the agenda of many parts of the global South. It is likely to generate violent conflict and, in some cases, to draw in core powers. While the forms of, and rhetorical justifications for, state-building interventions will probably change, there will be a regular demand for state-building operations for the foreseeable future. Although often intended to be short-term solutions to immediate problems, these interventions sometimes generate protracted efforts by the core power at societal transformation.

Core powers exhibit deep ambivalence toward such operations and often fail to implement the reform package in a sustained and coherent manner. Torn between pressures to promote global stability and reluctance to engage in costly and apparently futile interventions, core powers oscillate in their propensity to engage in military social engineering. As a result they repeatedly forget the lessons of such operations. The result is an expanded Vietnam syndrome: avowals of never again, followed by institutional amnesia, and then, after a sufficient lapse of time, a repeat of core power intervention.

The United States, like other core powers, has from time to time attempted to build states in other countries. Many of these efforts have been frustrating, and both political elites and mass publics have repeatedly vowed to refrain from similar undertakings in the future. It is the contention of this chapter that such interventions are difficult to do well and difficult to avoid. For the foreseeable future, great powers will be tempted to intervene in conflicts in the global South and will inevitably be drawn into protracted efforts at state building.

This chapter addresses four questions: (1) Why do core powers engage in social engineering in the global South? (2) Why are these interventions often unsatisfactory? (3) Why do core powers fail to learn much from these interventions? (4) Will such state-building interventions occur in the future?

State building is a difficult task, involving tensions and policy dilemmas. Core powers typically feel that they can invest only limited resources, and over time the resources available tend to decline as political costs mount. As it becomes more difficult to realize initial goals and expectations, efforts at social and political reform are abandoned in favor of a focus on strategic risk. However, far from generating a realistic reassessment of options, the increasing concern for strategic vulnerability tends to produce fantasy planning, precipitate withdrawal, and widespread disillusionment.

The hypothesis proposed here is that state-building military interventions by core states in the global South have been, and will continue to be, an infrequent and intermittent but important and salient aspect of warfare. The piece of the puzzle examined in this chapter is certainly not the most important global trend (that is, the disappearance of war in the core) and not a frequent phenomenon. It does, however, have the capacity to stir up political passion and lead to major foreign interventions with repercussions echoing through decades.

Supply and Demand

Core powers are under constant pressure to engage in protracted or repeated interventions and military occupations of client states and peripheral regions. There is a regular demand for, and a supply of, such interventions. These interventions sometimes generate efforts, more or less deliberate and coherent, at social and political reform and state building.

The demand stems from protracted problems in some parts of the global South. While there are reasons to be optimistic about long-run development trends for most countries, those with populations in the "the bottom billion" present serious challenges. Humanitarian concerns, such as famine and ethnic cleansing, draw core intervention. A range of issues, from ethnic conflict to environmental change, has the capacity to produce serious crises that may prompt intervention. In some regions (the Middle East, for example, or the former Soviet Union), conflicts over state boundaries and ethnic tensions combine with struggles for regional hegemony to produce a volatile cocktail of popular passion and state interest.

The reasons for foreign intervention stem from three main sources. The first and most obvious is humanitarian intervention: in many instances publics in the core will demand that something be done to reduce suffering among victimized populations. The second source is the perception that a security problem for the core is posed by instability in the global South. Not to be underestimated, the third is a sense of mission and national identity deriving from a more general vision of the nature of the core society, its place in history, and its role in the world.[1] All societies have a deep need for a sense of identity. One way this is often expressed is by naming the age: What sort of world do we live in, what are its drivers, where is it heading, what are the perils that we will confront, and what sorts of actions might we take? In some societies this is expressed in a belief that the core society has a duty to the rest of the world, has a particular mission, or should set a moral example.

The master frames and theories of grand strategy are not merely cognitive frames: they also assert the position of the actor within a larger moral universe. In the modern world, at least, war and peace are matters of moral purpose, not simply of state interest. They must be justified—and therefore conceptualized—in terms of identity and purpose in the world. In value terms, actors need a compelling narrative that deals with meaning, with matters of identity. This is why punditry is so important: narratives of danger must be morally compelling. Of course, once what appear to be purely cognitive issues become suffused with value and identity concerns, means-ends calculations become much more complicated, if not totally impossible. At the very least, the need for identity narratives drives out careful analysis. The purely cognitive task is hard enough; when it is embedded in the miasma of identity narratives everything becomes much more difficult. This is inevitable.

Sometimes cast as a secular notion of civilizing other parts of the world, or bringing the benefits of liberal democracy or modernization, and sometimes cast as a religious sense of national destiny, this sense of mission has been an important driver in the foreign policy of core nations.[2]

Occasionally there is a substantial degree of consensus about the meaning of our times and the role that one's society ought to play in the world; at other times, competing visions jostle each other. Decision makers in core states seldom agree entirely on the need for or the goals of such interventions. There is, rather, usually considerable debate and dissent. Sometimes the disagreements follow partisan lines; sometimes they cut across them. Identifying a pattern of core intervention is therefore difficult: what we see is ambivalence and oscillation. Policy can shift rapidly.

An additional dynamic comes into play: disappointment with the results of interventions leads to assertions that the core power will never repeat the mistake. Core powers, and particularly their military organizations, exhibit deep ambivalence toward such operations and often fail to follow through the reform agenda in a coherent manner. They repeatedly decide to forget the lessons of such operations.

Institutional learning is minimal. As a result, with the passage of time, the temptation to intervene resurges, generating a cyclical pattern.

Particularly in the aftermath of interventions that have not succeeded, there will be vows of never again, no more Vietnams. These political swings toward realism and neoisolationism may for a time reduce the likelihood of military intervention. However, the effect wears off, and new groups of decision makers, often under the illusion that they have an entirely new rationale or face a novel situation, embark on an intervention that turns out, whether intended or not, to entail large-scale efforts at state building. Indeed, the constant invocation of the never-again slogan is itself evidence of the failure to learn.

To illustrate the argument, this chapter briefly looks at three moments of history:

1. Early efforts at state building at the end of the nineteenth and the beginning of the twentieth century during the period of high colonialism. These interventions were justified in part by referring to the civilizing mission and in part by cultural assertions that the marker of great-power status was the possession of colonies. American acquisition of the Philippines during the Spanish-American War of 1898 serves as the example.
2. Cold War efforts at state building and counterinsurgency (with Vietnam under Ngo Dinh Diem as the illustrative case).
3. The post–Cold War interventions. These have come in two flavors: the liberal humanitarian interventions that characterized the 1990s, undertaken as peace operations to stop ethnic cleansing, and the neoconservative interventions in Iraq and Afghanistan that were the George W. Bush administration's operationalization of the global war on terror. The dominant public justification for these latter interventions was the security of the United States vis-à-vis a global terrorist threat. A subsidiary rationale, articulated principally by the U.S. deputy secretary of defense, Paul Wolfowitz, was that regime change—democratization—would create the basis for a peaceful world order.

This chapter argues that the similarities between these three historical periods of state building are more significant than the differences.

Imperial Tutelage: State Building in the Philippines

The period from roughly 1880 to World War I was dominated by a sense that core states had a duty to their colonies to guide them toward eventual independence. Imperial uplift was the theme, justified by theories of paternalistic and tutelary development. Colonialism was justified by the civilizing mission. The United States came late to colonialism, and its colonial territories were few. It was with the acquisition of colonial empire in 1898 in the aftermath of the Spanish-American War that deliberate efforts at state building overseas were undertaken. The Philippines were added to the emerging American empire almost as an afterthought. While a small group of policy makers that included Theodore Roosevelt, Henry Cabot Lodge, and Alfred Thayer Mahan[3] were inspired by dreams of empire, initially the U.S. intervention in the Philippines was aimed simply at preventing Spanish ships from leaving Manila to operate against the United States either in the Atlantic or around Cuba. Very quickly popular enthusiasm for taking the Philippines built up, and once troops were committed in large numbers, it was politically costly to turn back.

Once a decision was made to take over the Philippines from Spain (rather than to support national independence, for example, or hand the islands over to a European power), the United States had to make a series of decisions about the nature of the polity it had acquired. But first it had to suppress a popular insurgency. A large-scale counterinsurgency campaign dragged on until 1902, when the United States officially declared it at an end. Localized opposition continued for another decade before the islands were completely pacified.

From the very beginning the Philippines were a bitterly partisan issue, and the conduct of the counterinsurgency campaign generated vocal criticism in Congress and elsewhere. The military campaign was marked by widespread torture and brutality, and there was a public outcry as this became known in the United States. However, the reelection in 1900 of Republican president William McKinley set the seal on U.S. possession of the islands. The duty of the United

States was, in the words of President McKinley, to Christianize the population. American policy rested on three legs: education, political incorporation of Filipino elites, and economic development.

American educational goals were as much a cultural project as an instrumental one. Although much of the early emphasis in education was on vocational training, the importance of education in inculcating modern American attitudes was predominant.[4] At first, in an early example of a civic action program, U.S. soldiers were assigned to rural schools. Then, with considerable fanfare a large group of American teachers was brought to the archipelago. These teachers were endowed in the public imagination with a sort of missionary role in bringing enlightenment to the Philippines. Instruction was to be in English. The program seems to have been relatively effective in generating public support for the American occupation, though substantial numbers returned home early, disheartened or disillusioned with conditions in the Philippines.

The United States made early efforts to enable Filipino elites to exercise some political power once hostilities were largely over. The Filipinos were allowed to create a political party, the Federalista Party. The luminaries who formed the party were pro-American and opposed to raising the issue of eventual independence. This party provided a convenient mechanism for the articulation of elite views and a means to transmit American intentions to the elite. In 1901 American occupation authorities set up municipal councils and appointed local Filipinos to them. In 1907 they supervised the establishment of an elected assembly. The idea was that the Filipinos would be taught democracy from the bottom up, beginning with municipal government. This was an attempt to replicate what U.S. elites understood to be the practice of American town government. When Filipino nationalists petitioned to form a political party, Governor William Howard Taft refused: the limits of political debate were to be kept narrow.

The franchise was very restricted (about 2.5 percent of the population), and what rapidly emerged was negotiated control by local landowning elites. This took the form of patron-client relationships, which permeated the Philippine polity. Believing that the mass of Filipinos was not yet ready for democracy, the American rulers were happily complicit in the return of the Filipino landed classes to effective local power. The issue then became one of constantly renegotiating the relationship between the Filipino elite and American occupiers.

Filipino politicians were necessarily subject to cross-pressures and ambivalences. On the one hand, they needed to cooperate with the occupation authorities to maintain their positions, maintain access to resources, and generally further their interests. On the other hand, to retain mass legitimacy they needed to make nationalist gestures that would bring them into conflict with occupation authorities. The more farsighted of the American governors-general understood these dynamics, but they were nonetheless a constant source of irritation and tension in Filipino-American relations.

American efforts to promote good governance came to naught. This was inherent in the structure of politics established by the Americans. It basically removed fiscal responsibility from the Filipino political elite, creating dependency on the imperial power and the redistribution of spoils to ensure access to political office. It became patronage politics from top to bottom.

Under the early administrations of Governors-General Luke Wright and W. Cameron Forbes, the development of economic infrastructure, particularly roadbuilding, was seen as the main task of the colonial administration. American economic enterprises were given preference in many spheres, and this led to an emphasis on the export of primary products as the key to economic growth.

By perpetuating the dominance of local elites, the American colonizers avoided tackling major social problems such as inequitable land tenure. This foreclosed one path toward a more equitable style of development. Unequal land distribution made it difficult to translate the gains from trade into economic diversification and improved standards of living for the bulk of the population. Similarly, the relentless repression of organized labor by the security services of the colonial state suppressed the emergence of a radical labor movement in Filipino politics.

As Glenn Anthony May notes, Congress "never formulated a Philippine economic policy. Probably it was incapable of doing so. It was too large, too heterogeneous, to unwieldy, too poorly informed, and most of all, too little interested."[5] This meant policy incoherence and inadequate funding.[6] Politically, "the net effect of U.S. policies was not to implant popular government but to nurture a quasi-oligarchy."[7]

U.S. elites were often ambivalent about the Philippines; many of them saw continued U.S. possession of the archipelago as a burden. But exactly how—and when—to give the Philippines formal independence was a knotty political problem. From the perspective of American elites, a key issue was whether Filipinos were ready to run their own country. This was discussed primarily in racial terms. According to Edwin Kemmerer, economic advisor to the Philippines (and to various Latin American governments), "The average Filipino is proverbial for his lack of foresight and thrift. He is a creature of the present; his wants do not extend far into the future and even if they did he does not possess the capacity of sacrificing the trifling pleasure of the moment in order to satisfy them."[8] In 1921 the Forbes-Wood Commission was tasked to ascertain the fitness of Filipinos for independence. The commission concluded that "the time is not ripe."[9]

By the late 1920s voices advocating a slow, deliberate withdrawal had begun to exert influence in Congress. Debates about the pros and cons of retaining the islands were heavily marked by what Alfred W. McCoy has called "the politics of scandal."[10] Eventually, American legislators moved to withdraw from the Philippines. Among some groups was a clear sense that the Philippine experiment had been a costly and risky failure. Not only were the economic benefits dubious; the Philippines were a strategic liability. As Franklin Delano Roosevelt asserted during the 1934 congressional debates, "Let's get rid of the Philippines—that's the most important thing. Let's be frank about it."[11]

In 1932 legislation was initiated promising Philippine independence, but it failed to pass in both the United States and the Philippines. Eventually, the Filipino political elite negotiated a transfer of power toward independence with the passage of the Tydings-McDuffie Act of 1934 (promising independence in ten years' time) and the inauguration of a commonwealth the following year. In 1935, the Philippines achieved partial independence with Manuel Quezon as president and Sergio Osmeña as vice president. By the late 1930s a sense of crisis and anxiety had come to define the diplomatic reporting from Manila. There were two interrelated developments: the evolution of the Philippine polity toward a one-party state and the heavy shadow of approaching war with Japan.

While American efforts to develop the Philippine economy and to gradually transfer political power to Filipino elites were characterized by a measure of realism, efforts to cope with the military vulnerability of the Philippines were marked by a retreat into fantasy. It was this strategic irrationality that simultaneously brought disaster upon the people of the Philippines and produced national independence. And arguably, it was American support in the aftermath of World War II that enabled the Philippines to defeat a leftist insurgency and ensure that the Philippines would then develop in close economic dependency on the United States.

Strategy and Fantasy

Initially, American global visionaries had seen the Philippines as a stepping-stone to China: the islands would provide a jumping-off point for economic, religious, and military influence. But this was a double-edged sword: the Philippines would be an obvious target for any other power that sought to expand in Asia. At first, American strategists saw European colonialism as the major threat to the Philippines. By the end of the Russo-Japanese War of 1904–1905, it was apparent to many American strategists that Japan was a potential threat. The U.S. Naval War College began to war-game against Japan, and early versions of War Plan Orange (in which the adversary was Japan) were drawn up as early as 1906.[12] As time passed, Japan became the potential enemy against whom the Philippine Islands would need to be defended. This shift of strategic perception was swift and, as it turned out, accurate.

The Japanese were interested in the Philippines primarily because of their strategic importance: a major U.S. military base lying across their lines of communications to the resource-rich areas of the Dutch East Indies and Southeast Asia could not be left intact. It was obvious to all that the Philippine Islands would be hard to defend.

To a very considerable degree American war plans were fantasy documents. There were three sets of plans: the official War Plan Orange, which was continually being debated and revised; General

Douglas MacArthur's plan as military advisor to the Philippine government; and at the last minute, an air deterrence plan emanating from the Army Air Corps and eagerly taken up by President Roosevelt.

The central difficulty was that the Philippines were underdeveloped, and the Philippine Commonwealth could not on its own furnish a large modern defense force. This would entail a permanent reconnaissance out to sea. In turn, this meant a modern air force and a reasonably sized modern navy. There was no obvious way that a country as poor as the Philippines could afford such a force. Moreover, the need for coastal defenses and for a mobile armored reserve to strike at an invading army posed additional costs. Some of these costs could be borne by the U.S. taxpayer, but there were clear limits to this.

It was not simply the poverty of the Philippines but also the nature of its political system that militated against a sound assessment of the costs of defense. Filipino political elites were more concerned to ensure that government revenues went into lining their pockets, or those of their supporters, or into economic development rather than into defense.

Initially, two quite distinct concepts were developed, one by General MacArthur and another by the war planning departments in Washington. As the Philippines became a commonwealth, Quezon looked for a commander for its military forces. MacArthur was offered the position of advisor to the commonwealth.

MacArthur faced the difficulty of defending a vast archipelago with untrained forces and a nugatory budget. He began with the fact of underdevelopment and proposed a defense based on a large but low-tech military establishment. Taking Switzerland as his model, he proposed to create a large, militia-style army by means of compulsory service and a cycling of soldiers into a large reserve force. If the Philippines were to be attacked, trained reservists could be rapidly mobilized for local defense against invading forces. The invaders would be stopped at the beaches and thrown back into the sea. Simultaneously, swarms of PT boats would harass the invasion fleet, sinking troop transports. The combination of these measures would effectively deter invasion.

With typical bombast he announced:

> The keystone . . . is the trained citizen army. . . . [T]he Philippine Defense Plan is founded on the basic concepts of a free people. . . . There can be no doubt that so long as the population is imbued with the spirit of patriotism and determination it now so clearly evidences, a success, a brilliant success, is assured.[13]

His solution was magnificent in its invocation of the virtues of a citizen soldiery but totally impractical given the poverty of the Philippines.

American war planners took a different view. Beginning in 1906, they began to develop War Plan Orange. The central feature of the plan was a recognition that with the forces locally available, the Philippine Islands were basically indefensible. The most that could be accomplished would be to fight a delaying action, retreating to the mountainous peninsula of Bataan and the island fortress of Corregidor. U.S. forces would then have to hold out until the U.S. Navy could defeat the Imperial Japanese Navy, and the U.S. Army could return in strength and relieve the beleaguered garrisons on Bataan and Corregidor. If—as turned out to be the case—the Navy was in no position to rescue the garrison within a matter of months, the garrison would have to surrender to the invading forces. While War Plan Orange always contained considerable elements of fantasy, in contrast with MacArthur's scheme for the defense of the Philippine Islands, it was a model of sobriety.

Finally, as war with Japan loomed, Roosevelt and some key military decision makers pushed a new war plan. They sought to deter a Japanese attack on the Philippines by basing B-17 bombers in the islands. It was a case of too little, too late, and rather than deter a Japanese attack, the hasty U.S. buildup in the Philippines provoked an attack.

On the eve of the Japanese invasion, MacArthur was recalled to active duty and took charge of all military forces in the islands. He now shifted back—after initial efforts to prevent a Japanese landing—to the basics of War Plan Orange: the citadel defense of Bataan and Corregidor. Missing from the plan was the U.S. Navy. With the main battle fleet sunk at Pearl Harbor, it was in no condition to effect a speedy relief of the Philippines.

In the time frame allowed by the approach of World War II, it proved impossible to promote three key features of state building: social and political reform, independence, and national security. The reform initiatives were poorly conceived, failed to address questions of social equity, and were eventually dropped. As American legislators increasingly (and correctly) came to see the Philippines as a burden on the U.S. economy, they shifted to advocating national independence. The key decision came in 1934 when it was announced that after a ten-year transition period the Philippines would be granted full independence.

War, Occupation, Liberation, Independence, Counterinsurgency

Before the Philippines attained independence, they underwent the disaster of Japanese invasion and occupation and the costly liberation of the islands in 1944–1945. The Japanese occupation gave rise to the growth of a massive insurgency, some groups run by American officers who had gone to the hills when the Japanese invaded, other groups run by leftist parties, including the communists. One of these groups, the Huks, was later to emerge as a major threat to the newly independent government of the Philippines. American assistance in defeating the Huk insurgency of the early 1950s, discussed below, was an experience that proved formative for American notions of nation building and hearts-and-minds counterinsurgency tactics during the Cold War.

The immediate postwar tasks in the Philippines were daunting. There was a pressing need for immediate economic recovery. The economy was in shambles, Manila was in ruins, and much of the labor force was unemployed. The political status of the Philippines had to be determined and a decision made as to whether independence should be granted shortly or deferred until a more favorable moment. The police force was tainted by its collaboration with the Japanese, and the military needed to be rebuilt. As had been the case in much of the region where fighting and resistance had occurred, large parts of the local population were armed, organized, and politically mobilized. Some sort of disarmament and reorganization was essential.

Finally, longer-term goals of state building, democracy, and development needed to be addressed. In this context, U.S. policy makers wanted a rapid withdrawal. From the perspective of the American occupying power, everything boiled down to one question: getting the right man, and the right class of people, into power. The Americans wanted the Philippines run by someone they could trust, "our man in Manila." This was to be Manuel Roxas, first president of the independent Philippines.

The central implication of the Cold War context was a grand strategic interest on the part of the United States in the postwar development of the Philippines. It was important to ensure that the Philippines remained an anticommunist bastion in Asia.

In effect, this meant the return of the Filipino elite to political power, the retention of oligarchic dominance and patron-client politics, and an amnesty for collaborators. These decisions were to have long-term consequences for the country's political stability.

The Roxas government soon began to face a serious insurgency from the Huks. The Huks (Hukbalahap, short for Hukbong Bayan Laban sa Hapon, the Peoples' Anti-Japanese Army) had been formed in 1942 as the major anti-Japanese resistance group. In 1945 they retained the name Hukbalahap, but this now stood for Hukbong Magpalayang Bayan (People's Liberation Army).

In the elections of 1945, Luis Taruc and several other Huk leaders were elected to the Philippine congress but were prevented from taking their seats. This was a serious mistake on the part of the government: it persuaded many Filipinos that the electoral road was closed to reformers and that armed resistance to the regime was justifiable.

The historical roots of agrarian discontent go back to the period of Spanish colonialism. Landholdings were extremely concentrated before the Americans arrived. Most direct cultivators were sharecropping tenants. Benedict Kerkvliet argues that the terms of tenancy gradually eroded in the first part of the twentieth century.[14] As landlords modernized, they were less inclined to fulfill the paternalistic expectations associated with patron-client relations in the countryside.

By 1945, many peasants were deeply in debt and had grievances against landlords. The political power of landlords (over the police, judiciary, local government, etc.) was increasingly resented, particularly as the landlords were seen as withholding their part of the bargain: the relief of peasants in times of difficulty. Adding to this general sense that the patron-client underpinnings of tenant farming were eroding, increasing demographic pressure on the land was beginning to generate a sense of land scarcity.

The newly elected Roxas adopted an attitude of extreme hostility to the Huks and authorized the private armies of agrarian landlords to carry out Huk hunts. Initial operations by the Philippine Army against the Huks went badly. Poorly trained, poorly led, poorly paid, and poorly motivated troops were uninterested in engaging the Huks, and the better-informed Huks were able to easily slip past the cumbersome army sweeps. Stuck in static garrison duties, the army was seen by the local populace as parasitic. Underpaid and ill-disciplined troops routinely stole food from the local peasants. Troops had few positive social ties with the local peasants and increasingly saw them as aiding and abetting their enemy and eventually as the enemy itself.

The situation turned around with the selection in 1954 of Ramon Magsaysay as secretary of defense. Influenced by the Central Intelligence Agency's Edward Lansdale, Magsaysay shook up the army.[15] He removed senior officers and initiated intense inspections.[16]

Magsaysay instituted a number of other quite simple and effective changes. He raised the pay of ordinary soldiers so that they could purchase, rather than steal, food from the villagers. He cut through the red tape so that widows of soldiers killed in action would get an immediate cash advance. Later this was extended to soldiers wounded in action. These actions raised soldier morale considerably.

Army units were ordered to engage in active patrolling. Soldiers were instructed to enter villages in a friendly manner, to distribute candy to children, and in general, to treat the local population as fellow citizens, not as enemies. The army became more effective, did what its civilian leader wanted it to do, and ceased to alienate the local population.

Beyond army reform, as secretary of defense and later as president, Magsaysay instituted important reforms in the area of good governance. These were key to the success of the counterinsurgency campaign. Magsaysay decreed that all Filipino citizens could send a telegram of complaint to him, at little or no cost, and that it would be dealt with in twenty-four hours. A second reform provided army lawyers to the public as public defenders at little or no cost. Both of these were confidence-building measures in the central government. They showed that the state could be responsive to the grievances of its citizens.

Another important area where Magsaysay introduced major improvement in governance was to ensure that the election of 1951 would be a clean one, in marked contrast to the election of 1949. Napoleon Valeriano and Charles Bohannan note, "To all intents and purposes, the 1951 election sounded the death knell of the Hukbalahap movement."[17] The Philippines were now set on a path of dependent development as a firm bastion of U.S. Cold War efforts in Asia. The lesson drawn by many was that state building was an unnecessary and futile endeavor: the careful influence of a few key individuals and limited forms of American advice were more effective. It was a matter of carefully choosing our man in Manila.[18]

Cold War, Modernization, and Nation Building in Vietnam

The United States entered the Cold War with confidence in social engineering. The New Deal, the Tennessee Valley Authority, and World War II had convinced (at least some) elites that social change could be engineered and that the military was one of several agencies capable of undertaking that task. The occupations of both Germany and Japan seemed to underscore the point. The prevailing sensibility was one of high modernism.[19] The task was to remake that part of the world dominated by the West without falling back into the colonial habits of the past. This meant the creation of informal empire and an activist stance toward what was then known as nation building.

There were also, as Edward Miller has pointed out, many "low modernist" approaches to development planning, which drew more on the grassroots participatory dynamics of rural social engineering.[20] These, too, testified to the sense of American elites that they had the efficacy to guide social change, though they differed importantly in their belief that the ability of ordinary people to develop bottom-up initiatives was central.

The end of World War II left the world with a global Cold War between the two superpowers and vast regions of political turmoil and popular mobilization. American social engineering interventions were designed to thwart what was seen as communist subversion in the global South and to harness revolutionary sentiment for anticommunist purposes. By the early 1960s, influential thinkers picked up on a powerful strain of anticolonialism in the United States and advocated a policy of preemptive reform. This was embraced by President John F. Kennedy, with creation of the Peace Corps, establishment of the Alliance for Progress, and the search for a "Third Force" in Vietnam.[21] In this line of thinking, American goals in Vietnam were to build a South Vietnamese state that would enable that country to move toward modernity. For President Lyndon Johnson, that entailed another version of the Tennessee Valley Authority.

The vehicle chosen by the Americans was Ngo Dinh Diem, their "miracle man" in Vietnam. They chose to "sink or swim with Ngo Dinh Diem."[22] Not least of his merits was that he was a Christian. This made American policy makers comfortable. As Seth Jacobs notes, "More than in any other period in modern American history, policymakers in the 1950s tended to view events at home and abroad through a religious lens."[23]

The conceptual underpinnings of the reform strategy were provided by the emerging theory of modernization.[24] This theory suggested that deliberate intervention by core states would rapidly move the societies of the global South from tradition to modernity. Getting there, however, entailed risk. Communism was a disease of the transition. According to modernization theory, the effort to move from tradition to modernity invited efforts at communist revolution, which needed to be thwarted.

It followed that modernization should be pushed as rapidly as possible. During this critical stage, enhanced surveillance and repression might be required. Counterinsurgency, particularly in its hearts-and-minds incarnation, was thus an integral component of a broader modernization strategy.

Michael Latham has argued that "by the early 1900s, U.S. practices of imperial development reflected a volatile combination of reformist idealism and lethal coercion."[25] The key to understanding this part of the global dynamic is an appreciation of the tensions inherent in great-power intervention in the periphery.

The story of America's gradually increasing involvement in Vietnam has been told many times.[26] By 1954, if not earlier, the United

States had assumed responsibility for maintaining an independent South Vietnam in opposition to the communist North. Initially, this meant three things: acceptance of the division of Vietnam into two distinct states, North and South; a commitment to Diem—our man in Saigon; and a general policy of nation building. Diem's overall goal was to establish his personal rule and defeat or absorb his political rivals. Vietnam at this time was a patchwork quilt of contending sovereignties and quasisovereignties. The government in Saigon was a government in name only. Large parts of the country were ruled by private armies, linked either with the religious sects, organized crime, or regional warlords.

The failure of state building in Vietnam is often traced to American misperceptions about the nature of the conflict in Vietnam and about Diem himself. Certainly misconceptions abounded. American policy makers consistently failed to grasp the complexity of the war in Vietnam and pursued military strategies that were self-defeating. By framing the war in Vietnam in Cold War terms, the United States thwarted its earlier reformist impulses. By conceptualizing the conflict as arising from the cross-border aggression of a neighboring communist state (shades of Korea), the United States blinded itself to the internal civil war and peasant revolt. This was a hybrid war[27] and needed to be understood as such. Diem was seen as a neutralist, sympathetic to the United States, capable perhaps of embodying a third force, between right-wing dictatorship and communism. The corruption, clientelism, and administrative inefficiency of the Saigon regime were overlooked, as were the human rights abuses of the police. Both landlords and tax gatherers were seen by peasants as rapacious parasites, and the mandarin families that had been such important intermediaries of French rule were now seen as part of an illegitimate regime.

However, while there were indeed many misconceptions, more was at issue than simple cognitive error. American notions of state building were often at odds with those of Diem and his close associates, and this in itself made state building difficult. As had been the case in the Philippines, the American strategy of backing Diem was part of a long tradition of seeking a local strong man who would further American objectives. Unfortunately for the Americans, Diem did not behave in ways that they understood and approved of. This led to tensions and eventually to the Americans giving a green light for a military coup against Diem. Having cast aside their man in Saigon, the Americans had no one to replace him. The Vietnamese government was now run by a changing series of military dictators. The demands of the escalating war had driven out any serious effort at political and social reform.

Although it is usual to present Diem as a Catholic mandarin,[28] Miller and Phillip Catton point out that Diem had his own vision of a future Vietnam and his own program for state building.[29] The internal security imperative dominated: Diem needed to reestablish the authority of the central government over the territory it only nominally controlled. This required a complicated pattern of alliances and confrontations with, first, the religious sects, the Hao Hao and Cao Dai, and later with the Binh Xuyen mafia. Confrontation with the Buddhists followed. Throughout, Diem had to keep a wary eye on independent warlords and politicized elements of the army. Diem was successful at first, but the confrontation with the Buddhists in 1963 led to his downfall. Meanwhile, as Diem eliminated his rivals and as his police and security services became increasingly effective, the Vietminh insurgency intensified in 1959. It was this, together with military backing from the North, that eventually posed the security threat that brought down the South Vietnamese state. At this point strategic concerns swamped any reformist and state-building impulse.

Social engineering in Vietnam was attempted by Diem, mainly with his project of agrovilles and later with the Strategic Hamlet program. The Diem regime first attempted to settle families on virgin land. As pioneers, the settlers would clear the land and form new communities. They would be strong supporters of the regime. They would exhibit the hardy individualism and self-reliance that were thought to be central features of Vietnamese history. After this program had failed, the regime turned to the notion of strategic hamlets. These were a militarized version of the agrovilles, without the utopian aspects of reclaiming the agrarian frontier. The Strategic Hamlet program depended for its success on the ability of the political police to detect and neutralize National Liberation Front (NLF) agents and sympathizers within the hamlets. If it could not do this, then it would be impossible to keep the NLF out, and the entire program would be useless.

Efforts to concentrate the rural population into defended villages (agrovilles and strategic hamlets) were a fiasco. The security of the strategic hamlets required intense police surveillance and the creation of paramilitary organizations, which did little to win the hearts and minds of the rural population. Efforts at land distribution were minimal and were downplayed in favor of attempts to increase crop yields and reclaim unused land for farming.

There were multiple reasons for Diem's failure. The first, of course, is that the regime attempted to develop these programs in the midst of a war. The NLF did its best to undermine them. But the programs also failed for the sorts of reasons that state planning often fails: they were utopian and imposed on the population from the top down.[30] The implementation of these programs, moreover, was stymied by the nature of the Vietnamese state and by the political strategy of the Ngo family. Staffed to a large extent by the regime's political supporters, the state had little administrative capacity. It was riven by bureaucratic rivalries, and the family dictatorship aspect of the regime meant that there was little trust or scope for delegated authority.

The regime was fundamentally a patrimonial one. The key positions were held by members of Diem's family, and the regime relied heavily on a minority group—the Catholics—for its core political support. The administration was venal: government office was used for personal gain. One effect of this was that, particularly in rural areas, common people saw the central state as exploitative or in league with exploitative landowners. Corruption permeated the system and did much to undermine it.

The Diem regime also had difficult relations with its American sponsor. Both parties had in common the aim of creating a noncommunist state in southern Vietnam, but their vision of what this state would look like and, more importantly, the kinds of measures necessary to defeat the communists differed considerably. Managing these tensions proved difficult: the Ngos were prickly nationalists who feared that their powerful and rich ally would swamp them, and the Americans were often impatient with what they regarded as the inefficiencies of the Diem administration. Indeed, there was an undercurrent of skepticism about Diem within the American administration. President Dwight Eisenhower's special representative in Vietnam in 1954–1955, General Lawton Collins, quickly came to the view that Diem was a disaster and ought to be jettisoned. Despite an unwillingness in Washington to listen to him, "Collins stuck to his guns and . . . the wheels of a change in policy were already in motion" by 1955.[31]

Facing a series of challenges from political rivals, from politicoreligious sects, the Binh Xuyen, the communists, and the army, Diem's basic strategy was to force a showdown—on his own terms if possible—with one after another of these rival forces and eliminate it as a political force. In this, he followed a realistic, if risky, form of state building. The goal was to create the monopoly of the legitimate use of violence that all modern states strive for. To his American sponsors, brought up on a diet of pluralism and division of powers, this seemed both antidemocratic and self-defeating. They much preferred a strategy of compromise and accommodation. It is far from clear whether they were right in their criticisms of Diem.

Above all, of course, the demands of kinetic warfare against the NLF and the People's Army of Vietnam constantly churned Vietnamese society and made the pursuit of social engineering a chimera. The U.S. Army—and American decision makers more generally—misunderstood the nature of the war on which they had embarked. As a result, such schemes of social engineering that were attempted were poorly conceived, given insufficient support, and eventually ignored as the war became increasingly violent. The regime's efforts to win the allegiance (or acquiescence) of the population were now largely irrelevant and probably impossible.

By the late 1960s elites and publics in the United States had become bitterly divided. Even the principal architects of the war had been stripped of the illusion that they could achieve victory. The American polity began to cast about for a way to withdraw, and President Richard Nixon, with his Vietnamization policy, hit on the politically acceptable way to do this.

For many civilian analysts "no more Vietnams" meant acknowledging the limitations of social science as a guide to state building in foreign countries. American hubris was to be replaced with professional modesty. State building—and modernization theory in general—was to be abandoned as too difficult and too prone to blowback. By the last decades of the Cold War, state building was in broad disfavor and counterinsurgency had dropped off the radar.

After the Cold War

When the Cold War ended, pundits rushed to redefine the epoch. Eventually, three themes came to dominate: it was an age of globalization (though quite what this meant remained contested); terrorism was a new, persistent, and global threat; and armed conflict would now be of a new kind, deriving not from states but from substate actors, driven by ethnic conflict, resource scarcity, weak states, and religious fundamentalism, among other factors.

With the end of the Cold War, there was a reassessment of the situation of the global South. Market fundamentalism and a belief that democratization was the principal way forward to development had been building up since the late 1970s. These ideas were to triumph with the end of the Cold War. There were curious similarities with the theories of modernization that flourished earlier. Theories of globalization have, in fact, served largely to rejuvenate modernization theory. The emphasis has moved somewhat, to state failure rather than state building and to the risk of Islamic terrorism rather than communism. But the underlying logic has remained remarkably constant. Globalization presented both opportunity and threat. In American foreign policy, it took two forms: liberal humanitarian interventions and neoconservative efforts to promote markets and democracy.

Initially, American presidents displayed considerable caution in using military power. Their initial forays into humanitarian and decapitation operations were quite tentative. But as time passed and the notion that the United States was the sole superpower gained traction, the desire to use military power for larger purposes increased.

Under President George H. W. Bush, and then under President Bill Clinton, U.S. forces embarked on a range of peacekeeping and humanitarian missions, most notably in Somalia and the former Socialist Federal Republic of Yugoslavia. These efforts had mixed results. The social engineering that was attempted was quite limited and generally consisted in supervising ethnic segregation and power-sharing arrangements. Conservatives argued that they were a misuse of American military power. As Condoleezza Rice famously asserted in 2000, "We don't need to have the 82nd Airborne escorting kids to kindergarten."[32] The Democrats were to disagree.

During the George W. Bush years a number of quite distinct rationales were offered for U.S. intervention in other countries. Protecting citizens of the United States from international terrorism was perhaps the most salient, closely followed by assertions that intervention was needed to deal with an emerging threat of weapons of mass destruction being developed by rogue states. Opponents fastened on the prospects of oil riches to explain the invasion of Iraq in 2003. Mostly unnoticed, both then and now, was a neoconservative theory of social change. In a series of speeches in the months leading up to the invasion, Paul Wolfowitz articulated an ambitious program of imposed change.

At the regional level, this was seen as an exercise in social engineering: the problem with the Middle East and its offshoot, global jihadi terrorism, was the lack of democracy. The neoconservatives of the Bush administration expected to generate revolutionary change in the Middle East. In addition to the immediate goal of toppling Saddam Hussein, they hoped to unleash a cascade of democratic transformation and free-market regimes across the region. Remarkably, they expected these transformations to happen more or less automatically and very rapidly. They were, accordingly, surprised when they found themselves saddled with an economy that had collapsed, a state that did not function, and an insurgency that rapidly morphed into an ethnic civil war.

As these interventions rapidly confronted insurgencies and civil wars, the swift exit that had been imagined by neocon planners vanished. Americans in Afghanistan and Iraq found themselves doing something that they were not prepared for and that they felt ought not to be done: state building. Despite their adherence to a doctrine that markets are better social regulators than states, neocons found themselves engaged in social engineering at the national level.

It is not necessary to believe that the invasion of Iraq in 2003 was either justified or prudent to accept that President Bush and his closest advisors were committed to a project of radical social transformation in the Middle East. As Latham has suggested,[33] the historical moment when one might have expected the United States to lay down the burden of overseeing the rest of the world was at the end of the Cold War in 1990. This did not happen: there were a variety of military operations other than war, mainly of a peacekeeping nature. Then the attacks of September 11 led to the Bush administration declaring an existential war on terror, with no limitation in time or space. Although many are now vowing never again to do "stupid stuff,"[34] withdrawal from Afghanistan and Iraq has proved difficult. The threat of terrorism remains. Developing states continue to fight over populations and borders. Ethnic cleansing and sectarian conflict continue. When Syria descended into civil war, voices were raised to do something. Hunting Boko Haram, Joseph Kony, and other unsavory characters remains a popular agenda item with important segments of the American public. The demand for intervention remains active.

Strategic Considerations Always Trump Social Engineering

State-building interventions alter the strategic calculus. Quite apart from the intended outcomes, there are often unintended consequences. This was certainly the case for both the Philippines and Vietnam. Although it may to be too soon to say with any confidence, it certainly seems that the post–Cold War interventions have produced their share of unintended consequences.

There is seldom consensus on strategy.[35] There will always be debates about what are the interests of states and nations, and there will rarely be agreement on the means to further such interests. There will be voices both for and against intervention. Liberals will want to intervene in places and for reasons quite different from those that animate conservatives. Other liberals and other conservatives will oppose intervention. Looking for any large measure of consistency and coherence in the military policy of core states is a chimera.

Many analysts of strategy appear to believe that states routinely pursue carefully thought-out long-term strategies. In some instances this may be the case, but mostly this is an optical illusion resulting from the effort of historians to find some guiding thread in clashing interests and contending interpretations. Precisely because strategy is socially constructed, groups, organizations, and factions struggle over the framing of strategy. A consensus may emerge in some circumstances and may last for some while, but this is not invariably the case.

The Philippines were not seen as strategically relevant until they dropped into America's lap. Even then, they turned out to be a strategic liability. In Vietnam, the struggle to create a viable state in the southern part of the country collapsed with the Diem regime, and Vietnam spiraled into an orgy of violence, leaving state building in the mud. In both Iraq and Afghanistan early hopes have been cruelly dashed.

The Dynamics of Disillusionment

State building is difficult but not impossible. One problem is that expectations are too high: interventionist politicians and bureaucrats have false beliefs and naïve assumptions about how easy it is to change societies.

Libertarians of both right and left, Friedrich Hayek and James Scott, believe that the high modernism espoused by states invariably produces disastrous consequences. Markets or people (depending on the brand of libertarianism) usually produce better outcomes. Both agree that states do not know enough to engage in social engineering. However, states always believe that they are competent, including when one of their central dynamics is to reduce a complex reality to a simple slogan or formula.

Social engineering is frequently frustrating because it is difficult and because our understanding of how to promote it is rudimentary (and in the case of governments, often mistaken). A look at the literature on the history of state formation in the core should make the point: most sociologists believe that core state formation took centuries. Even allowing for the compression of time, to expect to create states in a matter of a decade or two is surely unrealistic. The process is complex, muddy, prone to reversal, and full of conflict.

The rhetoric of foreign military intervention is highly emotionally charged. This is inevitable, as political leaders and pressure groups seek to mobilize mass support for their policies. They must present intervention as a moral imperative, not merely as a matter of national interest. The tendency to frame foreign policy issues in moral terms is deeply rooted in the need of citizens to identify with their state. States become icons of identity, and identity is intrinsically a moral matter.

Foreign policy defines who we are. What is at issue is not simply cognitive framing but moral and symbolic framing. This invariably generates supercharged rhetoric. It leads to mistaken diagnosis and disillusionment when lofty goals are not met.

The Response of Military Organizations in the Core

After Vietnam, the U.S. military turned its back on counterinsurgency: with the exception of a small number of Special Forces soldiers with experience in Central America, the institutional army set about forgetting the lessons of counterinsurgency and had to reinvent the wheel in Iraq and Afghanistan. Something similar may well happen in the next decade or so. Policy makers will attempt to avoid another Iraq, and American military organizations will systematically unlearn the important lessons of recent wars. At some point in the future, however, reality will bite, and the military forces of the core will once again find themselves floundering in a quagmire. On this dimension, at least, war will not change its fundamental nature. What will change will be theories, understandings, and perceptions. What are in essence old wars of great-power intervention and state building will be seen as novel and unprecedented. The wheel will be reinvented.

There is a possible mitigating factor. As the U.S. military became engaged in Afghanistan and Iraq, it turned to anthropologists for a better understanding of what they were up against. This led, in turn, to a push for cross-cultural competence throughout the military. It remains to be seen whether the cultural turn in military doctrine will survive the winding down of the wars in Iraq and Afghanistan. In general, military organizations remain uncomfortable with counterinsurgency and state building. They are eager to forget the messy conflicts in which they have been engaged and return to the "real" business of soldiering: preparing for high-intensity warfare.[36] However, as wars between core powers vanish, the struggle to hold on to real war will become increasingly problematic.

Cycles

In the Philippines, the original enthusiasm for empire and social engineering rapidly began to wane as global economic crisis, congressional parsimony, and the looming threat of Japanese invasion all exerted pressure on American decision makers to find some way to withdraw from the Philippines. Congress increasingly came to see the islands as an economic burden and was influenced by sectors of American agriculture that saw Philippine exports to the United States as unwanted competition. The defense of the archipelago against an expansionist Japan posed an almost insoluble problem. The Philippines were poor and could not be expected to build a modern military organization without massive foreign assistance. But this was not forthcoming in the fiscally parsimonious 1920s and 1930s. With a small U.S. garrison, a limited commitment from the U.S. Navy, and an undertrained and under-resourced Philippine Army, any consideration of the defense of the Philippines inevitably generated fantasy planning.

Half a century later, as Washington policy makers increasingly came to appreciate that the war in Vietnam was turning out to be a costly political mistake, they sought ways to disengage. By 1973 American ground troops had departed, and in 1975 the regime in South Vietnam, which the United States had attempted to bolster against the communists since the early 1950s, collapsed. In the end, both Vietnam and the Philippines, for different reasons, turned out to be cases of strategic overstretch. This alone doomed the long-run prospects for social engineering.

Land reform was an obvious project for social engineering, and many aspects of modernization theory saw this as a key part of a broader development package. Land reform would simultaneously raise agricultural productivity and create a broad base of small farmers who would be the political base for center-left reforming governments. They would break the economic and political logjams imposed by the agrarian oligarchy. In neither the Philippines nor Vietnam did this happen. In both cases, the political costs were too high. In the Philippines two moments might have made a significant difference in land reform. The first was in the early days of the American occupation. The Americans might perhaps have built a base of support for their regime by embarking on an ambitious scheme of land and tenancy reform. They did not, opting instead to work through a revived form of political clientelism embedded in a form of tutelary democracy. The second opportunity for agrarian reform occurred in the late 1940s and early 1950s. The Philippines had just been granted independence but were faced with a rural insurgency in central Luzon, the Huks. The government passed some ameliorative legislation and resettled former guerrillas on state land but never undertook a widespread reform. In Vietnam, Diem relied heavily on rural elites for political support, and rewarding his Catholic followers was intended to come through the extension of agricultural land rather than through redistribution. Only later, during the period of military dictatorships that followed Diem's assassination, did the South Vietnamese government begin a widespread program of land redistribution. By then it was a case of too little, too late, and these programs had no discernible impact on the outcome of the war.

Efforts, both in the Philippines and Vietnam, to modernize the state, making it less corrupt and more effective in delivering services to its citizens, ran into difficulties early on. Vietnam's deeply rooted patrimonialism and the equally pervasive clientelism of the Philippines posed major obstacles to a structural reform. In addition, the economic policies sponsored by the core state were not particularly conducive to an egalitarian growth path that might have reduced political conflict. No doubt future historians will come to similar conclusions about Iraq and Afghanistan.

The history of failure in sponsored state building will, however, be no obstacle to repetition of the process. Both the demand and the supply of core state-building interventions will remain high. The need for state building in at least some regions of the periphery will remain obvious. As for the supply side of the equation, core states are committed to maintaining global stability and spreading the benefits of civilization. Important publics will continue to push for humanitarian interventions of one kind or another.

There will, of course, be intense political debate within the metropolis. There are always those who argue that involvement in the periphery does not advance national interests and may, indeed, hurt them by overextension. The balance between the cautious realists and the expansionists or interventionists has historically been a shifting one. The immovable object of turmoil in the periphery will meet the irresistible force of core urges to intervene.

A key driver of these dynamics is a sense of mission and perhaps of global responsibility. While sometimes realist arguments are advanced that stability in the periphery is a strategic interest of the core, in fact core powers can largely ignore what is happening in the periphery. The motives for intervention are frequently humanitarian. Sometimes interventions stem from ideological nostrums, like the magical power of markets to bring about political tranquility. This will not change.

Just as there will be pressures to intervene, so there will be forces making state building likely to fail. In addition to strategic overstretch, a combination of intractable reality, recalcitrant locals, and ideological simplification will combine to thwart state-building efforts. Disappointments will lead to vows never to repeat the process, and this means that any lessons learned will be rapidly buried. The way will then be open for a new generation of state builders to reinvent the wheel. Believing that they are dealing with a novel situation in a new and promising way, they will repeat past mistakes.

The final result, in both the Philippine and Vietnam cases, was that once initial enthusiasm and high expectations had disappeared, pressures to disengage became overwhelming. The process of disengagement from the Philippines was complicated: by the 1930s, American policy elites had begun to disengage, and no more Philippine adventures were contemplated. But before withdrawal could be effected, the Japanese invaded, postponing withdrawal until 1946. Defeat at the hands of the Japanese led to General MacArthur insisting on a reinvasion of the Philippines, before granting Philippine independence. Independence, of course, did not signify—as it had in Vietnam—a loss of American influence. The United States continued to be a protector of the Philippines and retained large military bases there. It actively assisted in the counterinsurgency campaign against the Huks. From a military perspective, the campaign against the Huks was seen as a success and as a model for other counterinsurgency operations. By the mid-1950s, far from deciding that there were to be no more Philippines, the lesson learned was that the Philippine model was of widespread relevance. What the campaign against the Huks apparently demonstrated was that social engineering was unnecessary. Cleaning up the political apparatus and forcing the army to become more professional (and less corrupt) were all that was required. In this sense, the campaign against the Huks in the Philippines offered support for modernization theory and for the grand nation-building dreams of the 1950s and 1960s. It showed how the United States could (partially) disengage and move toward a mutually satisfactory relationship without undertaking social engineering projects. In Vietnam military defeat led to the Vietnam syndrome: a widespread statement that Vietnam (or some aspect thereof) had been a mistake that was not to be repeated. For two decades or more, Vietnam was held up in political debate as a reason for the United States to avoid another quagmire. In Vietnam, disillusionment with Diem led to the military coup and assassination of 1963 and, in the ensuing political chaos that engulfed South Vietnam, to the Americanization of the war. Once this occurred, social engineering took a back seat to the big-unit war. By the time the Americans once again turned their attention to the war in the villages, it was too late. U.S. withdrawal and the subsequent collapse of the regime were followed by a chorus of voices proclaiming "no more Vietnams."

In both the Philippines and Vietnam, in the early stages of the conflict, some form of social engineering was attractive to the Americans and their local allies. Large schemes were developed, and responsible officials were generally optimistic about their ability to implement the schemes and about the impact such schemes would have on government legitimacy, pacification, and economic development. In both cases, organizational capacity was insufficient, popular enthusiasm and motivation were often lacking, and the programs failed to attain the goals set. Fundamentally, American decisions to work with existing local elites meant that there was massive resistance to structural reform. All actors, for diverse reasons, became disillusioned, and goals switched to more basic issues of regime survival and military defense. In the aftermath of what was widely seen as failure, metropolitan elites vowed never to repeat such mistakes.

For the two decades following American withdrawal from Vietnam, its global intervention was generally more circumspect. Inevitably, however, hubris and the urge to transform other societies by social engineering reemerged, this time under the banners of Clintonian humanitarianism and neoconservative theories about setting free market forces. Both are now in disfavor; but the old wars of state building may reappear. It would be a bold analyst who asserted that the cycle had finally come to an end.

Notes

Research for this chapter was funded in part by a grant from the Office of Naval Research, U.S. Department of the Navy, No. N00014-09-1-0557. The Department of the Navy is not responsible for the views expressed herein.

1. This is a social constructionist argument: it is not simply the objective problems of the global South that generate core intervention. It is also necessary that at least some key players in the core frame these problems in such a way as to make intervention a security concern for the core state.
2. A recent example is Andrew Preston, *Sword of the Spirit, Shield of Faith* (New York: Anchor, 2012).
3. Evan Thomas, *The War Lovers* (New York: Back Bay Books, 2010).
4. Cited in Peter Stanley, *A Nation in the Making: The Philippines and the United States, 1899–1921* (Cambridge, MA: Harvard University Press, 1974), 85.
5. Glenn Anthony May, *Social Engineering in the Philippines* (Westport, CT: Greenwood Press, 1980), 149.
6. Ibid., 146.
7. Ibid., 180.
8. Quoted in H. W. Brands, *Bound to Empire: The United States and the Philippines* (Oxford: Oxford University Press, 1992), 68.
9. Ibid., 120
10. Alfred W. McCoy, *Policing America's Empire: The United States, the Philippines, and the Rise of the Surveillance State* (Madison: University of Wisconsin Press, 2009), 97.
11. Quoted in Brands, *Bound to Empire*, 163.
12. Edward S. Miller, *War Plan Orange* (Annapolis, MD: Naval Institute Press, 1991).
13. Douglas MacArthur, *Report on National Defense in the Philippines* (Manila, Philippines: Bureau of Printing, 1936), 44–45.
14. Benedict Kerkvliet, *The Huk Rebellion* (Los Angeles: University of California Press, 1979).
15. Cecil Currey, *Edward Lansdale: The Unquiet American* (New York: Houghton Mifflin, 1998); Jonathan Nashel, *Edward Lansdale's Cold War* (Amherst: University of Massachusetts Press, 2005).
16. Napoleon Valeriano and Charles Bohannan, *Counter-Guerrilla Operations: The Philippine Experience* (Santa Barbara, CA: Praeger, 1962).
17. Ibid., 240.
18. The emphasis on limited engagement and advice and support was to reemerge in the aftermath of Vietnam, with American involvement in the wars in El Salvador and Nicaragua carefully calibrated to keep below the radar of public and congressional concern.
19. There was a great deal in common with both Stalinist and fascist attempts at social engineering, of course.
20. Edward Miller, *Misalliance: Ngo Dinh Diem, the United States, and the Fate of South Vietnam* (Cambridge, MA: Harvard University Press, 2013), 54–84.
21. Ibid., 32.
22. Seth Jacobs, *America's Miracle Man in Vietnam* (Durham, NC: Duke University Press, 2004), 2.

23. Ibid., 11.

24. Modernization theory has been ably dissected by a number of analysts, including Nils Gilman, *Mandarins of the Future* (Baltimore: Johns Hopkins University Press, 2003); Michael Latham, *Modernization as Ideology* (Chapel Hill: University of North Carolina Press, 2000); and Michael Latham, *The Right Kind of Revolution* (Ithaca, NY: Cornell University Press, 2011).

25. Latham, *The Right Kind of Revolution*, 18.

26. Mark Atwood Lawrence, *Assuming the Burden: Europe and the American Commitment to Vietnam* (Los Angeles: University of California Press, 2005); Frederik Logevall, *Embers of War: The Fall of an Empire and the Making of America's Vietnam* (New York: Random House, 2012); Kathryn Statler, *Replacing France: The Origins of American Intervention in Vietnam* (Lexington: University Press of Kentucky, 2007); Mark Atwood Lawrence and Frederik Logevall, eds., *The First Vietnam War: Colonial Conflict and Cold War Crisis* (Cambridge, MA: Harvard University Press, 2007).

27. Frank Hoffman, "Hybrid Warfare and Challenges," *Joint Force Quarterly* 52, no. 1 (2009): 34–39.

28. See, for example, Jacobs, *America's Miracle Man in Vietnam*.

29. Miller, *Misalliance*; Phillip Catton, *Diem's Final Failure* (Lawrence: Kansas University Press, 2002).

30. James Scott, *Seeing Like a State: How Certain Schemes to Improve the Human Condition Have Failed* (New Haven, CT: Yale University Press, 1999).

31. Catton, *Diem's Final Failure*, 10.

32. Michael R. Gordon, "The 2000 Campaign: The Military; Bush Would Stop U.S. Peacekeeping in Balkan Fights," *New York Times*, October 21, 2000, available at http://www.nytimes.com/2000/10/21/us/the-2000-campaign-the-military-bush-would-stop-us-peacekeeping-in-balkan-fights.html.

33. Latham, *The Right Kind of Revolution*.

34. "Is 'Don't Do Stupid Stuff' the Best Foreign Policy?" *New York Times*, September 2, 2014, available at http://www.nytimes.com/roomfordebate/2014/09/02/is-dont-do-stupid-stuff-the-best-foreign-policy-30.

35. Professional military strategists, like all professions, claim that they alone understand a body of esoteric knowledge. They claim that an objective technical competence and judgment is involved in discerning the strategic interests and concerns of a state and in devising appropriate courses of action to further such interests. Social constructionists—among whom I include myself—argue that interests are not objective but socially constructed.

36. David Ucko, *The New Counterinsurgency Era* (Washington, DC: Georgetown University Press, 2009).

4 /

A Crisis of Norms

Fighting Irregular Wars Well

Colonel C. Anthony Pfaff

After assuming control of Coalition forces in Afghanistan, Lieutenant General Stanley McChrystal adopted rules of engagement that imposed significant limits on collateral damage that American combatants could risk in the conduct of operations, even if it meant sometimes forgoing operations against insurgents. As he stated, "It's better to let a few insurgents escape than alienate the Afghan public by inflicting civilian casualties."[1] These rules reflected the belief, articulated in the U.S. Army and Marine Corps 2007 *Counterinsurgency Field Manual*, that collateral damage undermined efforts to draw popular support away from the Taliban. In this view, protecting the population would eventually lead to victory over insurgent forces.[2]

Shortly after the new rules went into effect, a Marine Corps company commander, viewing images sent from an unmanned aerial system, or drone, spotted four men planting what appeared to be an improvised explosive device (IED). He considered calling in an air strike, but when he noticed children playing nearby, he decided otherwise. Nonetheless, he had to do something: a convoy of U.S. Marines was going to pass by in a few hours. He continued to monitor the insurgents, and when the children appeared to wander away he decided to call in the strike. No sooner had he done that than the children wandered back into the strike zone. The children continued to play but never allowed themselves to get too far away from the insurgents. The captain concluded that the insurgents were using them as shields and called off the strike. After forty-five minutes, he decided not to attack the insurgents at all. Rather, the higher headquarters sent an ordnance disposal unit to the site to clear the IED before the convoy came through.[3]

It is a paradox of counterinsurgency that often the more force employed, the less effective it is. It is possible to learn from this the wrong lesson: never use force. However, it is also a fact of counterinsurgency—and irregular warfare in general—that employing no force is ineffective. In the preceding example, thankfully no Marines or children were harmed. Unfortunately, however, neither were any Taliban. As one Marine who was present reportedly stated, "We have to separate the insurgents from the people. . . . If we just bomb the hell out of everything, we'll have a hard time doing that."[4] On the other hand, bombing nothing or no one does not remove the threat to the population or the Marines.

In fact, avoiding risk of any collateral damage has generated a number of absurd situations. For example, while receiving mortar fire during an overnight mission, an Army sergeant reportedly requested supporting artillery fire—a 155 mm howitzer illumination round—so that his unit could better see the enemy's location. Despite illumination rounds not being designed to inflict casualties, higher headquarters rejected the request on the grounds that it could land in a civilian area and cause collateral damage.[5]

Of course, using force comes with its own problems. Kinetic operations (military attacks) may maximize risk to insurgents, but they also seem to increase, if not maximize, risk to counterinsurgents. This dynamic suggests another paradox of counterinsurgency: the more one protects one's force, typically by relying on overwhelming firepower that insurgents cannot match, the less secure these forces often are.[6] This insecurity arises not only because the increased risk of collateral damage can alienate the civilian population but also because force protection procedures tend to isolate combatants from that same population, giving insurgents space to conduct operations.

In fact, the emphasis on destroying insurgent elements by U.S. forces in 2003 to 2004 is often credited with the rise of the insurgency in Iraq and the ability of Al Qaeda in Iraq to gain support for its operations.[7] That experience drove the development of the counterinsurgency doctrine that generated the new rules of engagement. Further, the Israeli Defense Force (IDF) experienced similar frustrations during the 2000–2005 Intifada, when kinetic operations limited terrorist attacks but destroyed the Palestinian economy, which then pushed many Palestinians closer to Hamas.[8] In the 2014 conflict with Hamas in the Gaza strip, the IDF's reliance on kinetic operations carefully directed at Hamas targets seems to have empowered Hamas rather than degraded its ability to conduct operations.

This outcome arises because Hamas, like the Taliban, keeps civilians close when it operates. Hamas has consistently used schools, residential buildings, mosques, and hospitals as sites from which to launch rockets into Israel.[9] Further, many of these same civilians provide various kinds of support, from reporting on IDF activity to sheltering Hamas fighters. These practices not only make it more difficult for the IDF to target Hamas militants; they also build support for Hamas, especially when civilians are killed. These civilian deaths rally Palestinians, as well as members of the international community, to Hamas's cause, and local and international pressure on Israel to cease operations increases. Because Hamas benefits from Israeli strikes in this way, Israel's simply ceasing operations does not end the rocket attacks. Rather, this cycle typically ends when Israel has conceded something to Hamas that allows them to claim that the risks Hamas leaders took on Palestinians' behalf were worth it.[10] So, by putting Palestinians at risk, Hamas is better able to achieve its objectives.

These examples illustrate that neither protecting the population nor destroying insurgent capability will entail victory in irregular warfare—of which counterinsurgency is a type—no matter how well either or both are done. As Emile Simpson notes in his insightful *War from the Ground Up*, in irregular conflicts, military action, both "violent and non-violent, do[es] not have an intrinsic ability to influence people in a given way."[11] Further, these measures not only do not influence in a given way but often can be self-defeating. Such a dynamic confounds military planners' and doctrine writers' efforts to prescribe a single, coherent way of irregular warfare. The result is a crisis of norms, in which previously accepted answers to questions about the best and the right ways to fight irregular wars no longer work.

What is a conundrum for the counterinsurgent force can be a relatively straightforward path to victory for insurgents. Often, the more danger the population is in, the more support insurgents will have, even if they are the source of the danger. This dynamic has worked for Hamas and also holds in Iraq today. Islamic State of Iraq and the Levant (ISIL) insurgents operate close to civilians, thus ensuring collateral damage, and also implement an intimidation campaign that has resulted in the death, injury, and dislocation of thousands of persons. So despite the numerous claims that Iraqis in general, including Sunnis, oppose the kind of Salafism practiced by ISIL, its strength still grows. This dynamic has also been present in Afghanistan, where most civilian casualties have come at the hands of the Taliban.[12] Despite this, the Taliban also seems to grow stronger and Coalition forces weaker, the higher the civilian death toll climbs.[13]

Resolving these difficulties requires understanding irregular warfare, not simply as an alternative to regular warfare but as a response of weaker powers to stronger ones' reliance on regular means of war. Whereas stronger powers emphasized firepower and relied on professional militaries comprising infantry, armor, artillery, air, and naval forces designed to destroy similarly organized and equipped forces in open battle, irregular foes emphasized messaging and often relied on smaller, lightly armed organizations that pursued less direct means to convince their stronger foe to cease fighting.

Viewed this way, the source of the paradoxes becomes clearer: the ends, means, and norms of irregular warfare are not only different than those of regular warfare; they are designed to undermine and subvert regular warfare's ends, means, and norms. Clashes between regular and irregular forces can often be like different games played by two players on the same board, except that only one player seems to be aware of that fact and able to effectively take advantage of it. It should not be surprising, then, that irregular forces, who made the new rules, benefit in this new game.

Regular and Irregular Warfare

Given the success of irregular warfare over the last century, irregular fighters appear to be on to something that has so far eluded their regular opponents. What that thing seems to be is a set of rules (or norms) for combatting regular forces, while regular forces do not have sufficiently developed rules for combatting irregular forces. The resulting normative crisis has led to the absurd situations described above where combatants sometimes have to accept more risk than the insurgents they are fighting. The question to be answered then is what rules should the stronger side play by?

In general, warfare is governed by three competing norms: (1) accomplishing the mission, (2) minimizing harm to noncombatants, and (3) protecting the force. These norms, which hold for all types of warfare, force trade-offs because commanders have to decide where to place risk: accomplishing missions puts both combatants and noncombatants at risk, protecting the force puts mission accomplishment and noncombatants at risk, and minimizing harm to noncombatants puts combatants and the mission at risk. Where to accept or displace risk depends on the ends desired, the means available to achieve them, and the context in which those means are implemented.

Since the ends, means, and context of regular and irregular warfare are different, decisions about where to place risk should differ as well. Whereas regular forces aim to impose their will on the enemy, irregular forces seek to change their adversary's mind; whereas regular forces try to destroy similar forces in open battle, irregular forces fight indirectly, using force to message multiple audiences, building support for their cause; and whereas regular clashes are force-centric, with the population practicably separable from the outcome of the war, irregular warfare is population-centric, and the population determines that outcome.

Concepts

The national security community has struggled with how best to label conflicts like those in Iraq, Afghanistan, and Palestine. Terms such as "asymmetric," "fourth generation," and "unconventional" are unsatisfactory: all wars are asymmetric in some sense; generational labeling appears arbitrary when there is historical continuity; and as these kinds of conflicts become the norm, "unconventional" becomes less and less apt. Finding the right label for the kind of conflict discussed here is difficult, as the literature is rarely consistent. Words matter, however, and to develop policies and dedicate resources and personnel, those policy terms must be clear.

The British strategist Colin Gray argues that whatever war is, it is composed of, "more or less, but always to some degree, organized violence motivated by political considerations."[14] This service to policy, he argues, distinguishes war from other sorts of violence, such as recreational and criminal. In his view, this "nature" of war stays constant. What does change is its "character," which describes the specific ends and means associated with a particular style of fighting. As he states, "If you believe that different wars are examples of different types of political and social behavior, you invite serious error in understanding the continuities amidst the more or less obvious changes over time and in different contexts."[15]

Instead, he argues, from the strategist's perspective most wars are variants of what one would term "regular" or "irregular," where regular and irregular are distinguished by the choices distinctive security communities make regarding how they will compete and, when necessary, fight.[16] In this sense, these terms are relative. State actors can choose a number of ways to organize and equip their militaries, and different choices will generate different responses. This analysis, however, assumes that state actors will organize their militaries to respond to the most dangerous threat they perceive and that, from any particular national security perspective, those threats—and the responses to them—will be perceived as regular.

In this sense, the word "regular" is not so much descriptive of a particular character of war as it is simply the label given to how the United States—and most other state actors—have chosen to compete and fight. The threat was other state actors, and the response was to develop weapons and formations that were more powerful than those of their most likely opponents. This response led to a reliance on professional militaries[17] comprising varying combinations of troops, armor, aircraft, and ships, or "conventional forces" designed to confront other troops, armor, aircraft, and ships in open battle.[18] In these sorts of conflicts, the side with the most firepower and mobility typically won.[19]

In this context, what makes a war regular is not whether it remains the dominant form relevant to actors' experience; rather, as the dominant choice, it is the starting point to understand the irregular response. So unlike the term "conventional," which suggests frequency of use, "regular" in this sense conveys "original" and allows us to make sense of the response. In fact, two Chinese colonels, Qiao Liang and Wang Xiangsui, in their oft-cited book *Unrestricted Warfare*, argue that U.S. conventional success in the 1990s permanently transformed the character of war. The Chinese military would not be able to match U.S. conventional military capabilities and would have to find another way to defeat a stronger foe.[20] In response to superior U.S. conventional capabilities, the People's Liberation Army has adopted *shashoujian*, or "assassin's mace," which is an umbrella term for doctrinal development and acquisition of weapons systems aimed at enabling the "inferior" to defeat the "superior."[21] This doctrine relies on surprise as well as deceptive and unorthodox methods unknown to the adversary. The means employed under this doctrine—such as those I have described—are intended to deter, decapitate, blind, paralyze, or disintegrate enemy forces.

Confronting regular forces from a position of weakness suggests that irregular forces will comprise some combination of state and nonstate actors who oppose an existing state or international order for legitimacy. Means could include insurgency, terrorism, militant activity, and crime to "challenge the authority of states, challenge the rule of law, [and] use violence in unconventional, asymmetrical, and indiscriminate operations to achieve their aims."[22] Such groups, of course, may receive support from other states, as has been seen recently in Ukraine, Iraq, and Palestine, as proxy wars are paradigms of irregular conflict. However, what distinguish irregular from regular conflicts are the aims, a point I turn to below.

So if what counts as regular depends on the choices made about threats and how to respond, then from the perspective of the U.S. security community, regular warfare involves principally state actors fighting to impose or defend sovereignty over territory by employing similar conventional forces. By contrast, irregular warfare involves nonstate or weaker-state actors using principally unconventional means to compel the opponent to accept its interest. This interest could be support for a particular political order or acceptance of a particular policy objective, such as freedom of maritime navigation in disputed waters. This definition seems to conform with the U.S. Department of Defense's *Joint Publication 1-02*, which defines irregular warfare as "a violent struggle among state and non-state actors for legitimacy and influence over the relevant population(s). Irregular warfare favors indirect and asymmetric approaches, though it may employ the full range of military and other capacities in order to erode an adversary's power, influence, and will."[23]

Characteristics

While clearly not opposites in the sense that if there is one there is not the other, these definitions clearly indicate differences in these two forms of war that have important implications for how they *should* be fought. In this section, I discuss the different characters of regular and irregular warfare to show how their different ends, means, and contexts affect practice and the norms that govern warfare.

The ideal of regular warfare typically involves professional state-sponsored militaries destroying one another in open combat. It is closely identified with the Western, or American, way of war, which historian Victor Davis Hanson describes as reliant on "firepower and heavy defensive armament" and favoring frontal assault "hammer blows" to rapidly destroy enemy forces and withstand their response.[24] In this view, American thinking on war is driven by the idea of "enemy as existential threat," which must be defeated to preserve the kinds of individual freedoms that have shaped Western societies since the ancient Greeks.[25]

The idea of the enemy as existential threat is important to understanding the ends of regular warfare. An enemy's existence is incompatible with one's own. As the German philosopher Carl Schmitt said, "An enemy exists only when, at least potentially, one fighting collectivity of people confronts a similar collectivity. The enemy is solely the public enemy."[26] Enemies, at least in the technical sense described here, may be killed because they represent the existential negation of one's own community's way of life. All competition with the enemy is a zero-sum game: any gain for the enemy represents a loss for one's community. Thus all conflicts between enemies have the potential to result in war. When a conflict is between enemies, war is the only means of resolving it because only war seeks the existential negation of the adversary.[27]

Therefore, drawing on Carl von Clausewitz, the aim of regular forces fighting regular wars is to impose one's will on that enemy.[28] Doing so, however, does not necessarily entail breaking the enemy's will as much as it does eliminating his ability to resist one's own. Regular war is not so much about what the enemy or enemy population wants as much as it is its ability to resist the imposition of one's own ends. This is an important point. Imposing one's will on another does not mean that the other must adopt new goals. It simply means eliminating the other's ability to prevent the realization of one's own goals. Germany was able to occupy France in 1940 because it destroyed the French and British military forces defending it. There was no requirement for a referendum among the French population to justify the invasion, as there appeared to be when Russia seized Crimea from Ukraine in 2014.

Because the destruction of enemy forces entails the inability of the enemy to resist, regular warfare is characterized by the identification of military and political objectives: the defeat of the enemy's military forces entails the achievement of the political objective. This emphasis on destruction of enemy military capability yields an ontology of war comprising dichotomies.[29] In this kind of war, actors are either allies or enemies; actions in war entail resistance or surrender; and the end state of war is victory or defeat. It is true that enemies may fight to a stalemate, but such a state of affairs is not stable, representing only a suspension of hostilities until the sides decide to fight again. The state of war itself continues until one side has imposed its will on the other.

The logic of regular war is simple in expression but difficult in application. One side has imposed its will when the enemy no longer has the capacity to resist. One eliminates enemy capacity to resist by eliminating his combat capability faster than he can eliminate one's own. Doing this requires a strategy of annihilation—or at least attrition—that seeks a head-to-head battle aimed at destroying as much of the enemy's forces as possible, as well as his ability to generate new ones.

Regular warfare is characterized by the clash of forces that are similarly equipped and separable from the civilian population, a point I discuss in more detail later, and the assumption is that superior combat power entails victory. Although it is always possible to employ superior military force poorly—regular warfare does not ensure the stronger side's victory—this assumption drives how militaries equip, train, and employ their forces.

This assumption, however, fails in the context of irregular war. Examples such as the French in Indochina, the Americans in Vietnam, and the Russians in Afghanistan all suggest that a way of war that relies on strategies of annihilation and attrition has its limits. In fact, according to a survey of armed conflict from 1800 to 1998, significantly weaker adversaries defeated stronger ones approximately 30 percent of the time.[30] In particular, in "asymmetric conflicts," which the report defines as conflicts in which the force ratio between strong and weak actors is greater than five to one, not only were there a surprising number of weaker-side victories but the frequency of those victories increased over time. In fact, from 1950 to 1998, weaker actors in asymmetric conflicts won the majority—fifty-five out of ninety—of the conflicts surveyed.[31] These findings suggest that the coalitions that fought in Iraq and Afghanistan in the early twenty-first century should have anticipated a much more difficult time.

The reasons for these outcomes are complex and beyond the scope of this discussion to fully analyze. Some scholars point out that where there are significant gaps in relative force ratios, the weaker side will typically have a correspondingly higher resolve. Their logic is that the weaker side, knowing that it is overmatched, would resort to force only with significant, even existential, interests at stake. This dynamic entails greater political vulnerability for the stronger side—especially in democratic regimes, in which political leaders are more accountable to the public than in authoritarian regimes—since the gap in relative interest entails a gap in willingness to bear the burdens of war, providing the weaker side an effective strategy. Since the weaker side is not an existential threat, rather than defeat the stronger side's military forces, it needs only to impose a burden on those forces that the public in a democracy is unwilling to bear.[32]

While this explanation can account for the outcomes of some conflicts, such as the American efforts in Vietnam or Russians in

Afghanistan in the 1980s, it is insufficient to fully account for many of the other instances where asymmetric conflicts yielded asymmetric outcomes. As political scientist Ivan Arreguin-Toft notes, the assumption that willingness to assume greater risk entails greater resolve is not necessarily true. He notes, "A state that imagines itself 'leader of the free world,' for example, might rationally calculate that although the defeat of an ally in a distant civil war would be materially insignificant, its own survival as a free-world leader depends on a favorable outcome."[33] However, while relative interest may not fully account for political vulnerability, where that vulnerability exists, it does account for the stronger side's defeat.

A better predictor of stronger-side defeat is the interaction between the opposing sides' strategies. As Arreguin-Toft notes, in general, opponents can pursue direct or indirect strategies, in which direct strategies aim at destroying the opponent's forces and indirect strategies aim at breaking the opponent's will. In irregular warfare, while stronger sides may prefer to engage irregular combatants directly, those combatants' ability to avoid direct combat provides them a strategic advantage. To the extent that they can use that advantage and exploit the political vulnerability of the stronger, they can win. It is more difficult for the weaker side, however, when the stronger side employs its own indirect approach and targets the will of the insurgents and the populations who support them.[34]

According to this analysis, same-approach interactions, in which adversaries both adopt direct or indirect strategies, usually end in defeat for the weaker opponent. When both adopt a direct approach, the explanation is simple. The greater force ratio allows the stronger opponent to attrit the weaker adversary faster than it can replace its forces. When both adopt indirect approaches, the explanation is more complex. Given difficulties in applying a direct approach, the best strategy for the weaker side is to undertake an indirect strategy involving guerrilla or insurgent operations that aim at imposing a cost high enough to undermine support for the war. For the stronger, the best way to break the enemy's will is to target the weaker side's leadership as well as the noncombatant population in which its forces operate.

When the stronger side is willing to target the noncombatant population directly, they often—though not always—win.[35] This targeting of the civilian population does not have to be or, as I discuss later, should not be lethal or indiscriminate. Arreguin-Toft cites the Strategic Hamlet program in Vietnam, in which U.S. forces resettled Vietnamese civilians in fortified areas where they could not provide support to insurgent forces, and the Phoenix program, in which U.S. forces assassinated key Vietcong leaders, as successful examples of this approach.[36]

Further, successful counterinsurgencies do not always depend on the population preferring the measures taken. The British victory in the Boer War, for example, was widely attributed to British commander Lord Kitchener's three-pronged plan that included fencing off Boer land to restrict movement; incarcerating civilians to prevent their cooperation with Boer fighters; and burning farms to deny Boers food, shelter, and other supplies. These measures were successful; however, they also created a lot of collateral harm. Of the 160,000 Boers Kitchener placed in concentration camps, more than 20,000 died because of poor camp conditions.[37] The lesson here is that population-centric warfare does not necessarily require making the population happy.[38]

This view also helps explain the paradoxes described earlier. Applying a direct strategy where an indirect strategy would be more appropriate ensures a mismatch between ends and means when forces target the wrong persons and things in the wrong ways. So winning is not so much tied to finding the right balance between coercive force against insurgent targets and attractive hearts-and-minds efforts directed at the population, as the paradoxes suggest. Rather, it is finding the right indirect approach, which includes the use of force, to match that of the insurgent. Ideally, though not necessarily, such an approach would force insurgents to adopt a more direct strategy of their own, which would play to the strengths of regular forces.

In adopting an indirect strategy, stronger forces need to align force with aims. While explanations like those advanced by Arreguin-Toft may account for cases when significantly weaker actors defeat stronger ones, the political scientist Patricia Sullivan argues that they do not account for why the stronger actor makes poor strategic choices in the first place.[39] Accounting for this further difficulty requires analyzing war aims in terms of a range between those that can be attained by brute force and those that require enemy compliance with a particular policy objective. Brute force objectives, such as regime destruction or seizure of territory or resources, require the destruction of the enemy's capacity to resist. Coercive objectives, on the other hand, depend on the enemy's compliance, which requires changing the enemy's mind about what is in its interests.[40] The difficulty with this latter war aim is that it entails a gap between political and military objectives that does not exist with brute force objectives. With brute force objectives, the destruction of the enemy's military capabilities is sufficient to destroy that regime or seize the territory in question.[41] However, there is no necessary connection between attaining military objectives and coercing an adversary to change his behavior.

Thus, Sullivan argues, the more dependent on compliance is the objective, the more difficult it is "to translate that political objective into operational military objectives and to establish a link between battlefield military effectiveness and overall strategic success."[42] In fact, she notes, the use of military force to destroy an opponent's armed forces or inflict suffering on a civilian can be counterproductive if the goal is to compel that opposing government to change its policies or to increase long-term popular support among the civilian population.[43]

This point suggests that the wider the gap between military and political objectives, the greater the uncertainty regarding the success of any particular use of military force. Militaries are adept at discerning how much force is required to defeat another military, even given environmental complications such as terrain, weapons capabilities, and leadership.[44] However, discerning how much force is required to change people's minds is much more difficult. As Sullivan notes, "The amount of coercive leverage an actor can derive from a fixed amount of destructive capability is contingent on the target's willingness to absorb the costs imposed."[45] This can give irregular opponents more control of the conflict's outcome, since they get to decide how much tolerance for suffering they have. Additionally, an irregular opponent does not need to directly confront the opposing military force to determine that tolerance, allowing it to employ the indirect strategies articulated by Arreguin-Toft. As Sullivan notes, "It is difficult to predict costs or plan military strategies with any type of precision when success is dependent on reaching an inherently unknowable enemy breaking point."[46]

Because the weaker, irregular force is more in control of the outcome, it is able to inject uncertainty that undermines the stronger side. As Arreguin-Toft notes, when there is a delay between committing military force and realizing objectives, the stronger force tends to lose because its population typically has a higher expectation, corresponding to force strength, that victory will be swift. Furthermore, to avoid the higher costs a protracted conflict entails, the stronger side often "yields to the temptation" to employ "barbarism" to increase the cost of warfare to the enemy.[47] In fact, it was frustration from the slow progress in the Boer War that drove Kitchener to take the drastic measures that he did.[48]

What fills the gap between military and political objectives is the narrative through which various audiences perceive military and political acts, in which "audiences" can mean anyone who is affected by or can affect the outcome of the conflict. Unlike relatively polarized regular conflicts between enemies, the audiences in irregular war can possess very different interests and commitments and thus respond to the language of war in very different ways. In this context, force is not simply a destructive means to attrit the enemy; rather, it is the "language of war," as Simpson notes, that "links the use of force to political meaning."[49]

This does not suggest that the language of war is relevant only in irregular conflicts. However, in regular conflicts, the enemy relationship restricts the kinds of messages sent to simply "surrender."[50] Irregular conflicts, on the other hand, with varied audiences, actors, and commitments, allow for messaging of a greater variety, which opens up greater opportunities for additional messaging, even with adversarial elements.

Of course, messaging is meaningless unless it is aligned with war aims. As Sullivan's analysis suggests, the aims of irregular warfare are associated with changing specific audiences' minds to accept one side's interests. Typically, for the stronger-state power, those interests are acceptance of a particular government authority, as was the case with Afghanistan and Iraq. Realizing this aim requires the state authority to gain the cooperation of the population and prevent them from cooperating with insurgents. For the weaker power it requires compelling the stronger to change its policy toward the weaker actor, as in Palestine and Ukraine. Realizing those aims also requires the cooperation of the stronger power to accommodate the weaker side's interests.[51]

This suggests that military forces in irregular conflicts will have to interact with fighters and the population among which they operate in complex ways—much more complex than is required in regular wars. Consider, for example, the battle at Ganjgal in 2009, where Corporal Dakota Meyer and Captain William Swenson earned the Medal of Honor. They were responding to a request for humanitarian assistance by village elders. However, as their unit approached the village, approximately sixty insurgents, with whom the villagers were apparently cooperating, opened fire. When Meyer's unit called for fire support, higher headquarters denied it out of fear of causing civilian casualties. In this instance, the desire to win hearts and minds not only lured U.S. forces into a trap but kept them in it.[52] While it seems apparent that whatever hearts-and-minds objectives U.S. forces had got muddled, it is more extraordinary they had such objectives at all. It is hard to imagine the mayor of a German town still under Nazi control convincing an American infantry platoon to provide critical services, like electricity. The request just would not be taken seriously. More importantly, their providing that assistance would not have appreciably affected the outcome of the war.[53]

The shift in emphasis from forcefully imposing one's will to compelling acceptance of one's interest impacts critical features of regular war.[54] First, unlike enemies, an irregular opponent attempting to compel a stronger power to accept its interests is not an existential threat. Compelling a political entity to accept one's interests, in fact, necessarily entails the continued existence of that entity. This point does not mean that irregular adversaries cannot be existential threats. Hollywood has already imagined for us terrorist groups armed with weapons of mass destruction wishing to impose some apocalyptical political agenda. I am not concerned with those kinds of threats here.[55]

However, this does entail that opponents, like insurgents in Iraq and Afghanistan, who do not seek to fundamentally alter U.S. governance or the American way of life but rather seek accommodation for their existence as a political entity are better considered as a class apart from enemies. For this reason, I refer to irregular opponents as "adversaries" in order to account for this qualitative difference between the enemies of regular war.

From this shift in ends emerges a view of war that expands on Clausewitz, changing war's scope. Ally and enemy are replaced by adversaries, collaborators, supporters, and sympathizers; resistance and surrender are replaced by acceptance and rejection; and victory and defeat are replaced by success and failure. Regarding this last point, in contrast to regular warfare, the end of an irregular conflict is not always clear, as there is seldom a formal surrender ceremony and, as the French experienced in Algeria, a relatively thorough destruction of insurgent forces does not always translate into winning.[56] Even if no one is left on one side to participate in a surrender ceremony, the other side can still lose. Further, "ally" and "adversary" do not refer simply to states but to substate and nonstate organizations as well. Additionally, such conflicts are not zero sum. If one side can achieve its interests by benefiting the adversary, or some subgroup within the adversaries' community, then it should, all other things being equal, do so.[57] Further, by shifting the emphasis of the war effort from imposing one's will to changing others' minds, the emphasis shifts from the destruction of enemy forces to engaging the indigenous population.

In irregular conflict, the enemy operates within the civilian population and depends on that population for shelter, food, medical assistance, finances, and other types of support. This relationship entails collaboration on the part of some, if not all, of that population, though it does not follow that this support is given willingly: the ability of the enemy to extract it determines its strength. Hamas, for example, while certainly enjoying support from segments of the Palestinian population, has a history of routinely placing rocket and mortar positions near schools, residences, and other civilian sites to exploit the resulting collateral damage, as well as forcibly moving civilians into areas where the Israelis are expected to attack.[58] This relationship draws the civilian population, willing or not, into the class of things that are logically inseparable from war fighting, making them necessary, if not legitimate, targets of war.

Norms of War: Winning, Minimizing Harm, and Protecting the Force

The increase in complexity of interaction between military forces and the population in irregular conflicts suggests an increase in the complexity of the norms that govern those conflicts. When the aim is the destruction of enemy forces and the enemy is practicably separable from the civilian population, decisions about risk are relatively straightforward: maximize risk to the enemy, minimize risk to friends, and minimize harm to noncombatants. However, when the aim is to obtain cooperation with a particular order or authority and the enemy is not separable from the civilian population, then minimizing risk to adversaries, increasing risk to friends, and exposing at least some noncombatants to harm may occasionally make moral sense. In the next section I examine how the different character of irregular warfare influences how the norms associated with war fighting inform decisions about risk.

Winning, Accomplishing Missions, and Military Necessity

Military ethics begins with the imperative to win, and winning requires combatants to undertake actions that achieve objectives that collectively lead to the realization of the political objectives military force is intended to attain. That winning is an ethical imperative may at first seem counterintuitive. We normally experience ethical norms as restraints on what we want to do, and winning wars is something any good combatant wants, even if he may not want to do some of the things required. Even Michael Walzer, author of *Just and Unjust Wars*, believes that military necessity is simply an amoral concession to the exigencies of war rather than an ethical imperative itself.[59] However, the logic is fairly simple: if the cause is just, achieving it maximizes the good. Thus, actions that lead to victory or avoid defeat in pursuit of a just cause are not just permissible but obligatory.[60]

Judgments about military necessity are, therefore, judgments about what sorts of persons and things contribute to enemy combat power and are thus liable to be targeted. What makes them liable is that they are practicably inseparable from the opponent's capability to wage war. Thus, rather than judgments about probability and risk, judgments about military necessity are more judgments about who and what combatants may permissibly target.[61] However, in determining what sorts of persons and things may be targeted, military necessity also tells us what sorts of risks combatants must take and what the limits on those risks are.[62]

What counts as necessary is a function of the aims and means of war. In regular warfare, the aim is to impose one's will on the enemy and the means are forces capable of destroying corresponding enemy combat power. So in regular war, military necessity obligates combatants to fight, which requires them to take risks. Since the enemy in regular warfare is an existential threat, combatants must accept existential risk. Put simply, to be a soldier means being obligated to risk one's life to destroy the enemy.

Military necessity also places limits on this risk. While individuals may be subject to existential risk, the military units they compose are not. As a collective, combatants represent the ability of the state to wage war. Since military necessity, to be a moral obligation, presumes a just cause, preserving the capability to continue fighting and to win places limits on the risks combatants must take before they can seek to displace it elsewhere. As a result, combatants are not required to take so much risk that the mission will fail or that they will not be able to carry on the war effort. As Walzer notes, the "limits of risk are fixed, then, roughly at that point where any further risk taking would almost certainly doom the military venture or make it so costly it could not be repeated."[63] Thus, when those limits apply, combatants may displace risk onto noncombatants. While, as I discuss shortly, those risks also have limits, this means that while combatants may not intentionally target civilians, when targeting enemy forces, foreseen though unintended harm to civilians is permitted.

The aims and means of irregular warfare also shape how military necessity restrains and enables combatant activity. The ethically appropriate aim of irregular war is to establish or preserve a just order in the face of an armed challenge. While realizing these aims certainly includes attritting insurgent forces, it also includes efforts to gain the population's cooperation with the institutions associated with that order. Destruction of the irregular opponent's forces will typically be permissible; however, it may not always be necessary, especially if there are nonlethal alternatives to obtain the required cooperation between insurgents, their supporters, and the government. In the gap between military and political objectives, force can only message; it cannot guarantee the aim on its own. This creates moral space where combatants are required to consider nonlethal measures against adversaries.

In the case of irregular conflict, military necessity involves a combination of coercive and attractive measures and thus both can be obligated given a particular context. While combatants may still be obligated to take risks, combatants must make choices regarding coercive measures intended to attrit insurgent capabilities and attractive measures intended to engender popular cooperation. While both are permitted and adjudicated by whatever most likely will work, the nonexistential nature of the threat suggests that combatants are permitted to not only prioritize less risky, even if less effective, alternatives but also forgo risky alternatives in favor of less risky ones.

So even if the children had not been present, forgoing risky alternatives suggests that the Marine captain in the opening example was right to send an explosive ordnance disposal (EOD) team to dispose of the IED rather than, for example, sending a rifle squad to kill the insurgents, assuming the risk to the EOD team was lower, even though doing so allowed the insurgents to get away. Thus, in irregular conflicts, combatants are permitted to take short-term risks during missions that favor developing less risky alternatives. This permission, however, has its limits. Irregular warfare is still warfare. Success, in part at least, does depend on killing or at least detaining irregular adversaries, especially their leadership.

Minimizing Harms to Noncombatants: Proportionality and Discrimination

Traditionally, the moral justification for minimizing harm to noncombatants begins with the principle that it is wrong to harm innocents, who have a right to life. In the case of war, innocence is more a function of one's relationship to war fighting than personal responsibility borne for the war. However, while there are many ways to argue this point, I stipulate that something is morally significant about persons that must be respected—even in war—unless the person has done something to make them liable to harm.[64] This respect requires combatants to take noncombatant rights into account when conducting operations. Taking those rights into account requires proportional and discriminate applications of force.

Whereas military necessity rules out only gratuitous harm, proportionality places a further utilitarian restriction on the use of force: when using force does more good than harm.[65] Military necessity also specifies the goods against which to measure proportionality. To the extent an act contributes to the war's aims, it counts as a good. To the extent it undermines those aims or results in violating the requirements of discrimination, which I cover in the next section, it counts as harm. So proportionality applies only to operations required by military necessity and overrides that necessity if the harm caused by the operation outweighs the good.[66]

Proportionality is a principle that is easy to articulate but difficult to apply. As a rule, it is a forward-looking criterion that requires combatants to calculate future outcomes on the basis of incomplete information. However, comprehensive and accurate information regarding an opponent's situation is rare in war, and even when the information is available, it is almost impossible to know that until after one has acted on it. Furthermore, it can be difficult to accurately calculate the impact of a particular operation on the opponent's situation. Even in the relatively straightforward calculations of attrition warfare, combatants cannot be certain that any given operation will achieve the desired outcome.[67]

Thus, for regular war, in general, the destruction of things practicably inseparable from war fighting, like enemy combat forces and munitions factories, count as goods, and the destruction of persons and things practicably separable, like civilians and their residences, are harms. In irregular warfare, proportionality calculations are much more complicated simply because more persons and things are practicably inseparable from war fighting. However, whereas the good in regular warfare is the destruction of enemy forces, the good in irregular warfare is a just civil order. While eliminating challenges to that order certainly counts as a good, so does building cooperation. Undermining that cooperation would count as a harm.

Proportionality rules out only certain acts, not certain kinds of acts. So while proportionality can determine that a particular act of murder is wrong, it cannot determine that murder is wrong. As long as the good outweighs the harm, any kind of act could be justified. As a result there is additional moral pressure to further restrain the use of force. This additional restraint ties into the presumption that noncombatants have a right to life that requires combatants to discriminate between the kinds of targets they may legitimately engage and the means they may legitimately employ. Military necessity sets apart those persons and things that are liable to an attack. Being essential to the enemy's combat capability, for the most part, assimilates persons and things into the class of legitimate targets of war. Those persons and things that are not so assimilated may not be targeted.

This notion of practicable inseparability from war fighting allows us to distinguish between military and civilian as well as combatant and noncombatant. To be practicably inseparable from war fighting, someone or something must have no other function than a war-fighting function. For example, enemy combatants are legitimate targets because the function of a combatant is to wage war. Farms and hospitals, even when they are used to feed and heal enemy forces, are not. The reason they are not is that these facilities and the people who operate them are engaged in activities required even when there is no war. Thus, argues Walzer, the relevant distinction to make is between not what contributes to the war effort and what does not but rather what an enemy soldier needs to fight and what he needs to live. Since the latter is required by combatants and noncombatants alike, it cannot be assimilated into the class of things associated with war fighting and thus should not be targeted.[68]

In regular war, the practicable separability of combatants and noncombatants means that noncombatants are immune from attack. This noncombatant immunity protects them whether they are complicit in the crime of war or represent a particular threat. It is simply the case that the population of nonthreatening, noncomplicit individuals is found in the civilian population, not the military one. Populations are considered nonthreatening and noncomplicit precisely because of the way regular wars are fought: clashes of similar forces in open battle. Thus civilians and civilian activities, for the most part, are irrelevant to the outcome of those battles and are thus immune from attack. Because they are immune from attack combatants must not only avoid targeting them; they must always choose the course of action that minimizes noncombatant risk of harm, even if it means taking additional risks to themselves.[69]

The reason combatants must take additional risks to preserve friendly noncombatants' lives is because they receive training, equipment, and other resources to reduce their risk when fighting. Since enemy and friendly noncombatants have the same right to life, it follows that soldiers must accept some additional risk to preserve enemy noncombatant lives as well.[70] Being immune from attack, however, does not mean being immune from harm. Because risk to combatants is limited by military necessity, they are permitted to displace some of that risk onto noncombatants. So while combatants may not intentionally target noncombatants, they can target enemy combatants knowing, but not intending, that some noncombatants will be harmed.

Discrimination in irregular warfare is much more complex. Because civilians are practicably inseparable from irregular forces' combat capability, they are presumably assimilated into the class of persons and things that may be targeted. What makes them necessary targets is that the militarily weaker side could not fight without their support and the militarily stronger side cannot win if it does not undermine the weaker side's effort and build support of its own. As retired British general Rupert Smith notes in *The Utility of Force*, in such conflicts, "the loyalties, attitudes, and quality of life of the people do not simply impact the outcome; they determine it."[71]

However, as the philosopher Thomas Nagel argues, a person may be subjected to hostile treatment only by virtue of the threat that person represents, as "hostility or aggression should be directed at its true object."[72] It is that object that enables and constrains permission to harm. As Nagel states, "In an altercation with a taxi driver over an excessive fare, it is inappropriate to taunt him about his accent, flatten one of his tires, or smear chewing gum on his windshield."[73] It is inappropriate because his accent, tires, and windshield are not relevant to the excessive fare. So in the case of irregular warfare, not every adversary employs lethal force and thus does not warrant a lethal response.

Three points about irregular warfare follow from Nagel's analysis. First, combatants may target only whatever it is about the noncombatant adversary that contributes to the adversaries' cause. So if the loyalty, attitude, or quality of life impedes counterinsurgent success, then only those loyalties, attitudes, or quality of life are legitimate targets. Second, it follows from the first point that only those means that specifically address barriers to success may legitimately be used. Third, "noncombatant adversary" needs to be introduced as a third category of persons relevant for normative analysis in irregular war.

Combatant adversaries, because they use lethal force, are subject to lethal force. However, noncombatant adversaries are those collaborators, supporters, and sympathizers who make up the "sea" Mao Tse-tung said was necessary for all successful insurgencies. Because they do not employ lethal force, they are not subject to it; however, that does not tell us much about what measures are permissible.[74]

However, given noncombatant adversaries' positive relationship with combatant adversaries' ability to conduct operations, it seems unreasonable to prohibit friendly combatants from engaging combatant adversaries just because noncombatant adversaries are present and may be harmed. This suggests that while noncombatant adversaries may not be targeted directly, they may be subject to the same kind of unforeseen harms noncombatants are subject to in regular war. Noncombatants in regular war are subject to those harms, not because they deserve it, but because and only if there is no other way to attack the enemy. The analogy is civilians involved in munitions factories. When they are in the act of making munitions, they may be attacked. This is because of the direct and lethal connection munitions have with war fighting. However, their residences may not be attacked, because when they are in those residences, their connection to lethal activity is no longer direct. However, to the extent their residences are proximate to the factory, they are subject to collateral harm since the factory itself is always a legitimate target because its connection to war fighting is always direct.

So when noncombatant adversarial activity is so integrated into combatant adversaries' lethal activity, friendly combatants may engage the combatant adversaries, using means that risk foreseen but unintended harms to noncombatant adversaries. Prudence associated with military necessity will limit this permission, since exposing noncombatants to harm, even if not intended, is no way to win goodwill. This point suggests that when conducting offensive operations in which noncombatants are at risk of harm, combatants must use the least force possible to accomplish the mission and, where possible, prefer nonlethal to lethal means. So, for example, the Marine in the opening example got it right by preferring to send an EOD team to dispose of the IED as opposed to attacking the individuals apparently planting it, since at the time he could not be certain of their identity, their intent, or the extent of collateral damage an attack could cause. This choice does not mean he should not find a way to kill or capture the insurgents; however, it does mean he will have to find a way that does not endanger children, even if those children were witting participants. It also suggests that in defensive operations, whatever means that reduce the harm to friendly combatants under attack would be justified, even if they entailed foreseen harm to noncombatants, as long as they conformed to the principles of proportionality and discrimination. I take this point up again in the next section.

Of course, not everyone—or even almost everyone—in a population experiencing irregular combat is a noncombatant adversary. Many are simply noncombatants, who have no connection to the adversaries' combat capabilities. Thus, while it may make some moral sense to expose noncombatant adversaries to the risk of foreseen harms because something about their activities warrants it, this is not the case for nonadversarial noncombatants. If the purpose of force is to preserve and maintain order, then subjecting noncombatants to collateral harm is no way to accomplish that. For example, the International Crisis Group found that the shelling of targets in Fallujah by Iraqi Security Forces (ISF), just after the city was taken by ISIL insurgents, was sufficiently indiscriminate to establish a vicious circle: to address the danger the ISF represented to them, civilians in Fallujah were forced to rely more on ISIL, and this reliance further justified the use of indiscriminate force by the ISF.[75] That is no way to defeat an insurgency.

Because some sense of order exists—weak and fragile though it may be—the environment of irregular war can be analogous to civil societies where a law enforcement ethic applies. Police, too, are charged with preserving order and thus are prohibited from engaging even violent criminals in a way that places civilians at risk of foreseen, though unintended, harm. However, police are permitted to engage in activities that expose civilians to unforeseen risks. For example, police are permitted to engage in a high-speed car chase even though there is a chance they may lose control of the car and harm a civilian. Here the harm is not foreseen, only the risk. If this permission holds for law enforcement in conditions of peace, it certainly holds in conditions associated with irregular warfare.

Protecting the Force

While military necessity obligates soldiers to take risks, a broader view of military ethics must also consider the obligations commanders have to preserve their soldiers' lives and well-being. Such measures have both utilitarian and deontic aspects to their justification. From the perspective of military necessity, commanders are obligated to preserve their forces so they may continue the fight. From a deontic perspective, these soldiers are human beings with their own rights to life and liberty.[76]

From a utilitarian perspective, the obligation to accomplish missions, coupled with the obvious fact that more combat capability means success is more likely, is an obligation to preserve that capability, which obviously includes soldiers' lives. This is especially true in regular warfare, in which one wins by destroying the enemy faster than the enemy can destroy one's forces.

In general, this is where the regular view ends. It might be enough to say that military necessity provides sufficient justification to protect combatants' lives. As noted several times above, the practical needs of war always give commanders reasons to preserve, if not privilege, the lives of their combatants over the lives of noncombatants. While considerations of proportionality and discrimination may not always permit them to do so, it does not seem commanders need much encouragement—or ethical argument—to be motivated to carefully guard their combatants' lives.

Even Walzer argues there is nothing more that needs to be said. In his view, to gain the right to kill, combatants give up their right to life.[77] Because they have given up their right to life, their deaths are not unjust. Since all combatants give their rights up in this way, they are moral equals and no combatant—just or unjust—does anything wrong by killing another combatant. This view, however, leads to a number of counterintuitive conclusions. First, it suggests that the deaths of just combatants at the hands of unjust combatants are themselves morally permissible. But it would be odd to say, for example, that a German soldier in World War II justly killed a French soldier since, by taking up arms, the French soldier had forfeited his right to life.[78] Doing so would suggest that someone can forfeit one's rights by virtue of responding to the unjust act of another. On that basis alone it seems reasonable to reject Walzer's position.

Second, such a view treats soldiers as mere means, which is aptly illustrated in Stanley Kubrik's movie *Paths to Glory*. In the film, French generals order a large-scale attack against an impregnable German fortress. Predictably, the attack fails. But rather than take responsibility for giving bad orders, the generals prosecute selected troops for cowardice and have them executed. If military necessity was the only justification for preserving soldiers' lives, then the French generals did nothing wrong. As long as they believed the fortress was a legitimate military target and as long as they believed these random executions would motivate increased élan for the next attack, they might plausibly argue their actions were, in fact, obligated.

However, the state's obligation to its citizens does not stop when they become combatants. Lieutenant General (ret.) James Dubik, arguing against Walzer's view, claims that combatants, like friendly and enemy noncombatants, retain their right to life even in war. He argues that if the rights combatants are defending are indeed natural rights, it is morally wrong to force them to give up those rights. He points out that while some combatants certainly choose to fight, it is the enemy's aggressive act that places them in circumstances where they are compelled to make that choice. Because the enemy has coerced them onto the battlefield, they do not give up their right to life.[79] This point is echoed by the Israeli philosophers Asa Kasher and Amos Yadlin, who argue that a "combatant is a citizen in uniform . . . [whose] life is as precious as the life of anyone else."[80] The state may be justified in putting their lives at risk because of its obligation to protect all citizens, but the obligation to protect those soldiers remains.

Thus the imperative of force protection suggests that policies requiring combatants to accept all the risk associated with an operation are morally questionable. So the Army sergeant in the second example should have at least received the illumination rounds he requested. Additionally, Meyer's platoon should have received the kind of fire support they needed to break contact with the Taliban with the least risk to his unit. In regular or irregular conflicts, combatants should have available to them all means to respond to enemy contact with the least risk to themselves. While military necessity, as well as discrimination, prohibits combatants from committing foreseen, though unintended, harms to noncombatants when conducting offensive operations against insurgents, the principle of force protection suggests that when soldiers are under fire and less lethal means are not available or would not be effective, then foreseen though unintended harms to noncombatants may be permissible, subject to proportionality and discrimination.

Conclusion

On the basis of this analysis, the source of the counterinsurgency paradoxes becomes clear, as does the way to resolve the crisis of norms they create. Because the weaker adversary controls the outcome in irregular war in ways that the stronger does not, these paradoxes are better understood as false dilemmas. Combatants do not have to choose between using greater force and assuming greater risk. Rather, they need to employ indirect strategies aimed at eliminating insurgent leadership and building cooperation among the population to separate insurgents, terrorists, and criminals that challenge a just authority. It is well beyond the scope of this discussion to fully articulate what sorts of measures would make up a successful indirect strategy; however, I suggest using a normative framework to evaluate those measures.

Whereas in regular war military necessity obligates the destruction of enemy forces, it does not do so in irregular war. What it obligates is using coercive and attractive measures to build cooperation regarding a particular order. Whereas in regular war proportionality and discrimination permit foreseen but unintended harms to noncombatants, in irregular warfare they permit only unforeseen and unintended risk to noncombatants unless friendly combatants are in contact with armed adversaries. Whereas in regular warfare combatants assume an unlimited liability to risk, mitigated only by military necessity that permits displacing some risk to noncombatants, in irregular war combatants do not assume an unlimited liability to risk, and policies that prevent them from displacing risk either to the mission or to noncombatants are immoral.

While military ethics is experienced as a restraint on the use of force, any ethic that undermines the ability to successfully wage war is nonsensical. The interaction of these norms addresses that concern and provides a framework for a comprehensive military ethic that assimilates practical aspects of military necessity with the moral aspects of noncombatant immunity and force protection. Therefore, as the Marine captain who called off the drone strike demonstrated, the balance of military necessity, harm minimization, and force protection means combatants should, in general, prefer nonlethal to lethal means; however, that preference ends when combatants can reasonably discern insurgents apart from the population and engage them. This latter point suggests that combatants should seek such opportunities out, even if it means engaging in practices the population itself may not prefer. Despite strong prohibitions on harming noncombatants, even noncombatant adversaries, the examples of the Army sergeant and Meyer's platoon illustrate that irregular warfare is still war and that some circumstances override such prohibitions: when friendly lives are at stake.

Irregular warfare is a response of the weaker to the reliance of the stronger on firepower to fight war. However, while war's nature may be eternal, its character is ever changing. Given irregular warfare's success, we are seeing state and nonstate actors turn to hybrid warfare as a means to impose will—as the Russians did when seizing Crimea—and change policy, as appears to be the case with China's assassin's mace, which is more suited to getting an adversary to accept China's interests than imposing its will. As these sorts of conflicts become conventional, it will be worth examining the application of the norms of warfare to discern best practices associated with fighting hybrid wars and the likely response.

Notes

The views expressed in this chapter are the author's and not necessarily those of the U.S. government.

1. Michael M. Phillips, "Civilians in Cross Hairs Slow Troops," *Wall Street Journal*, February 22, 2010, available at http://www.wsj.com/articles/SB20001424052748704751304575079741450028512.
2. U.S. Army and Marine Corps, *Counterinsurgency Field Manual* (Chicago: University of Chicago Press, 2007), 48.
3. Phillips, "Civilians in Cross Hairs Slow Troops."
4. Ibid.
5. George Will, "Afghan Agonies," *New York Post*, June 21, 2010, http://nypost.com/2010/06/21/afghan-agonies.
6. U.S. Army and Marine Corps, *Counterinsurgency Field Manual*, 48.
7. Peter Mansoor, "Army," in *Understanding Counterinsurgency*, ed. Thomas Rid and Thomas Keaney (London: Routledge, 2010), 77.

8. Mansoor, "Army," 79.

9. Anne Barnard and Jodi Rudoren, "Israel Says That Hamas Uses Civilian Shields, Reviving Debate," *New York Times*, July 23, 2014, available at http://www.nytimes.com/2014/07/24/world/middleeast/israel-says-hamas-is-using-civilians-as-shields-in-gaza.html.

10. Ibid.

11. Emile Simpson, *War from the Ground Up* (New York: Oxford Press, 2013), 154.

12. Dexter Filkins, "'09 Deadliest Year for Afghans, U.N. Says," *New York Times*, January 14, 2010, p. 6. According to Filkins's report, 2,412 civilians were reported killed in 2009, a jump of 14 percent from 2008. Another 3,566 were wounded. What is significant about 2009 is that this was the first year that most of the deaths—two-thirds (1,630)—were caused by Taliban. The number of civilian deaths caused by NATO forces fell by 28 percent, to 596, because of the restrictive measures imposed by McChrystal. Most of those deaths (359) were caused by air strikes. However, the Taliban appeared to have suffered sufficient backlash from their practices that they published a code of ethics emphasizing population protection. See "Taliban Issues Code of Conduct," *Al Jazeera*, July 28, 2009, available at http://www.aljazeera.com/news/asia/2009/07/20097278348124813.html.

13. Filkins, "'09 Deadliest Year for Afghans."

14. Colin S. Gray, "War: Continuity in Change, and Change in Continuity," *Parameters* 40, no. 2 (2010): 6.

15. Ibid. 7.

16. Ibid.

17. For the purposes of this discussion, "professional" refers only to the relative institutionalization of a military organized force. In this sense, a professional military has an institutional infrastructure for planning, recruiting, training, and acquisitions that irregular forces typically do not have.

18. Victor Davis Hanson, *The Western Way of War* (Berkeley: University of California Press, 2009), 10.

19. However, "most dangerous" does not mean "most common," because nothing about regular war entails its frequency relative to other kinds of war. It simply specifies a kind of threat around which states have chosen to organize a military response. As a result, conventional forces may often find themselves in unconventional roles, as is evident in Ukraine's struggle against Russian-supported separatists. So we can have a conception of "regular" that does not entail how often such wars occur.

20. Qiao Liang and Wang Xiangsui, *Unrestricted Warfare* (Beijing: People's Liberation Army Literature and Arts, 1999), 24. An English translation of the book is available at http://www.c4i.org/unrestricted.pdf.

21. Andrew F. Krepinevich Jr., "The Pentagon's Wasting Assets," *Foreign Affairs* 88, no. 4 (2009), available at https://www.foreignaffairs.com/articles/united-states/2009-07-01/pentagons-wasting-assets. See also Jason E. Bruzdzinski, "Demystifying Shashoujian: China's 'Assassin's Mace' Concept," in *Civil Military Change in China: Elites, Institutes, and Ideas after the 16th Party Congress*, ed. Andrew Scobell and Larry Wortzel (Carlisle, PA: Strategic Studies Institute, 2004), 309–364.

22. Richard H. Shultz and Andrea J. Drew, *Insurgents, Terrorists, and Militias: The Warriors of Contemporary Combat* (New York: Columbia University Press, 2006), 10.

23. Department of the Army, *TRADOC Pam 525-3-0: The Army Capstone Concept; Operational Adaptability: Operating under Conditions of Uncertainty and Complexity in an Era of Persistent Conflict, 2016–2028* (Washington, DC: Department of the Army, 2009), 48. See also U.S. Department of Defense, *Joint Publication 1-02: Department of Defense Dictionary of Military and Associated Terms* (Washington, DC: Department of Defense, 2010), 123, available at http://www.dtic.mil/doctrine/new_pubs/jp1_02.pdf.

24. Hanson, *The Western Way of War*, 9–10. According to Hanson, "American armed forces in recent wars have sacrificed mobility, maneuver, grace, if you will, on the battlefield in exchange for the stark, direct assault, of frontal attack against the main force of the enemy and the opportunity to strike him down—all in the hope of decisive military victory on the battlefield" (10).

25. Tony Pfaff, *Resolving Ethical Challenges in an Era of Persistent Conflict* (Carlisle, PA: Strategic Studies Institute, 2011), 10. For a discussion of Hanson's views, see Martin Shaw, *The New Western Way of War: Risk Transfer and Its Crisis in Iraq* (Cambridge, UK: Polity Press, 2005), 30.

26. Carl Schmitt, *The Concept of the Political*, trans. George Schwab (Chicago: University of Chicago Press, 1996), 28–29.

27. Ibid., 33.

28. Carl von Clausewitz, *On War* (Middlesex, UK: Penguin Books, 1968), 101.

29. Peter Paret, "Education, Politics, and War in the Life of Clausewitz," *Journal of the History of Ideas* 29, no. 3 (1968): 395. That Clausewitz would express himself this way is not surprising, given his exposure to philosophers like Immanuel Kant, who emphasized the role of will, and G.W.F. Hegel, who emphasized a dialectical process to reasoning. But Paret notes that while Clausewitz was certainly exposed to Hegel and Kant, he did not seem to directly apply their insights to his own views of politics and war. Thus one must be careful when inferring philosophical influences on his writing. While it seems clear that Clausewitz was not deliberately employing a Hegelian dialectic to construct a full-blown philosophy, I suggest he was influenced by the methods of great thinkers, like Kant and Hegel, in the way he approached the subject.

30. Ivan Arreguin-Toft, "How the Weak Win Wars: A Theory of Asymmetric Conflict," *International Security* 26, no. 1 (2001): 96–97.

31. Ibid., 97.

32. Ibid., 98.

33. Ibid.

34. Ibid., 100–105.

35. Ibid., 107–109.

36. Ibid., 117. Arreguin-Toft notes that, despite the success of these strategies, the United States still lost. A more unambiguous example of successful targeting of insurgent populations is British use of concentration camps in the Boer War, discussed later.

37. Martin Dugard, "Farmers at Arms: The First Modern Insurgency—and How the British Crushed It," *Military History* 27, no. 1 (2010): 33.

38. The point here may also be that otherwise legitimate measures may be implemented immorally. For example, Saudi Arabia implemented a similar approach near the Yemeni border during conflict with the al Houthis in 2009.

A Saudi Army officer told me that approximately three hundred thousand persons were moved into more or less permanent housing outside their villages. At the same time, the Saudi government poured millions of dollars into the local economy to alleviate any economic hardships as a result of the fighting.

39. Patricia L. Sullivan, "War Aims and War Outcomes: Why Powerful States Lose Limited Wars," *Journal of Conflict Resolution* 51, no. 5 (2007): 499.

40. Ibid., 504.

41. Ibid., 505.

42. Ibid.

43. Ibid.

44. Ibid., 506.

45. Ibid., 507.

46. Ibid.

47. Arreguin-Toft, "How the Weak Win Wars," 105.

48. Dugard, "Farmers at Arms."

49. Simpson, *War from the Ground Up*, 15.

50. If there were more space for communication between enemies, then it would not be an enemy relationship, which is characterized by very limited conditions for cooperation.

51. For example, in Ukraine, separatists seek autonomy from the Ukraine government, while Russia wants to change the Ukrainian government's policy of favoring ties with Western Europe. Regarding Palestine, Palestinians in general have a number of particular complaints against Israel, including expansion of Israeli settlements and restrictions on movement. While these complaints are in part over territory, these movements are seeking accommodation to administer territory they already live in, not seize it.

52. Jonathan S. Landay, "'We're Pinned Down': Four U.S. Marines Die in Afghan Ambush," *McClatchy*, September 8, 2009, available at http://www.mcclatchydc.com/2009/09/08/75036/were-pinned-down-4-us-marines.html. See also Bing West, "The Afghan Rescue Mission behind Today's Medal of Honor," *Wall Street Journal*, September 15, 2011, available at http://www.wsj.com/articles/SB10001424053111904353504576566302974342330.

53. This is a conceptual not practical point. It is, of course, possible that a set of circumstances arises in which opposing commanders can agree to allow humanitarian acts. Doing so, however, would require setting aside aspects of the enemy relationship that are practicably difficult to routinely set aside in war. Such instances are rare for a reason.

54. Qiao and Wang, *Unrestricted Warfare*, 140.

55. My concern is not necessarily the stated intent of a particular group. Hamas, for example, represents itself as an existential threat to Israel; however, it does not possess the capability to realize that intent. What is of concern is the intent behind the irregular forces' operations, which in these cases seem to be better described as changes in policy.

56. Ivan Eland, *The Failure of Counterinsurgency: Why Hearts and Minds Are Seldom Won* (Santa Barbara, CA: Praeger, 2013), 29.

57. Benefiting combatant adversaries should not feature largely in any irregular strategy. Rather, the point is that in irregular warfare space exists for such measures as practically, if not morally, necessary in an overall strategy.

58. Human Rights Watch, "Rockets from Gaza," August 6, 2009, available at https://www.hrw.org/report/2009/08/06/rockets-gaza/harm-civilians-palestinian-armed-groups-rocket-attacks.

59. Michael Walzer, *Just and Unjust Wars: A Moral Argument with Historical Illustrations*, 2nd ed. (New York: Basic Books, 1992), 131.

60. Ibid., 130–132. This logic contrariwise entails that if the combatants' cause is not just, then the harms they commit are not justified. This is a matter of some controversy, and it is beyond the scope of this discussion to fully explore it. However, for the purposes of this discussion, I assume a just cause for the stronger side. This logic does not entail that, absent a just cause, combatants should not observe the rules of war; rather, the point is that even if they did, the harms would not be justified.

61. Ibid., 144.

62. While military necessity is generally viewed as a justification to kill, it also restricts combatants to refrain from any killing that does not contribute to victory. Thus, military necessity also entails a prohibition against excessive or gratuitous harm. As long as the use of force serves a military purpose, military necessity is not concerned with the value of the military objective relative to collateral harms caused. So this restriction is not as burdensome as the principle of proportionality; however, given the history of wanton destruction in war, it does impose an important moral burden on war fighters.

63. Walzer, *Just and Unjust Wars*, 157.

64. Brian Orend, *War and International Justice: A Kantian Perspective* (Ontario: Wilfrid Laurier Press, 2000), 18–19. See also Walzer, *Just and Unjust Wars*, 135–137.

65. All theories of utility emphasize the maximization of some good and the minimization of some harm. They disagree, however, on what exactly the good is. For the father of utility theory, Jeremy Bentham, it was pleasure. For John Stuart Mill, it was happiness. Others consider it interest or well-being. In the context of just-war thinking, the good is understood as victory for the just side, which is the best outcome of any war, even for the enemy. See William Shaw, "The Consequentialist Perspective," in *Contemporary Debates in Moral Theory*, ed. James Dreier (Malden, MA: Blackwell, 2006), 10.

66. Michael L. Gross, *Moral Dilemmas of Modern War: Torture, Assassination, and Blackmail in an Age of Asymmetric Conflict* (Cambridge: Cambridge University Press, 2010), 162.

67. Often decisions about proportionality are made before combat in how a military chooses to train and equip its forces to fight. For example, to avoid noncombatant casualties the United States has sought to develop increasingly precise weapons that not only better acquire legitimate targets but minimize the harm done when engaged. So for the Marine captain seeking to target the insurgents planting the IED, proportionality was assumed, in part, in the means he had to target them.

68. Walzer, *Just and Unjust Wars*, 146.

69. Ibid., 157–159.

70. Ibid., 157. To illustrate his point, Walzer describes the example of French bomber pilots in World War II who were charged with attacking Nazi facilities in France. If they bombed from a high altitude, they were safer but less accurate.

If they bombed from a lower altitude, they were more accurate but in greater danger. In the end they chose to take risk and bomb from a lower altitude than the optimally safe one because they recognized that they had an obligation to their citizens to minimize the harm done to them, even if it meant taking extra risks. Walzer's point is that noncombatants on any side enjoy the same rights, so considerations one would feel obligated to apply to one's own citizens would also apply to enemy noncombatants.

71. Rupert Smith, *The Utility of Force: The Act of War in the Modern World* (New York: Knopf, 2005), 281.

72. Thomas Nagel, "War and Massacre," *Philosophy and Public Affairs* 1 (1972), republished in Timothy L. Challans, ed., *A Modern Symposium on Military Ethics as an Introduction to Philosophy* (Cincinnati, OH: Thomson Learning, 2000), 192–193.

73. Nagel, "War and Massacre," 193.

74. Mao Tse-tung, *On Guerrilla Warfare*, trans. Samuel B. Griffith (Chicago: University of Illinois Press, 1961), 93.

75. International Crisis Group, "Iraq: Fallujah's Faustian Bargain," Middle East Report Number 150, April 28, 2014, p. 5.

76. James M. Dubik, "Human Rights, Command Responsibility, and Walzer's Just War Theory," *Philosophy and Public Affairs* 11, no. 4 (1982): 355. Dubik rejects Walzer's conception that soldiers give up their right to life in order to gain the right to kill. He argues that if rights to life and liberty are indeed natural, then soldiers—or anyone for that matter—cannot give them up.

77. Walzer, *Just and Unjust Wars*, 145–146.

78. Jeff McMahan, "On the Moral Equality of Combatants," *Journal of Political Philosophy* 14, no. 4 (2006): 385.

79. Dubik, "Human Rights," 363–364.

80. Asa Kasher and Amos Yadlin, "Military Ethics of Fighting Terror: An Israeli Perspective," *Journal of Military Ethics* 4, no. 1 (2005): 19–20.

5 /

Searching for Red and Blue in the Fog of Gray

The Development and Deployment of U.S. Military Biometrics in Iraq and Afghanistan

TRAVIS R. HALL

> A war fighter needs to know one of three things: Do I let him go? Keep him? Or shoot him on the spot?
>
> —Anh Duong, inventor of the "bunker buster," on her efforts to create mobile biometrics for the U.S. military[1]

> A majority of [population groups] are gray [have unclear motives].
>
> —U.S. Central Command, "Concept of Operations for Biometrics in the US Central Command AOR"

American Forces Network Afghanistan, the U.S. military–produced television station for troopers abroad, airs an internal news program called *Freedom Watch Update*. In a report that aired August 28, 2007, a unit of the 101st Airborne sets up a checkpoint to biometrically capture the local populace as they move through their daily routines. Images show young U.S. soldiers setting up the checkpoint and fingerprinting and photographing smiling or stoic Afghans with what looks like a large, clunky camera.[2] The broadcast is a compilation of selected highlights depicting what the U.S. military wants biometrics to be. Yet a little digging online yields a more complicated picture, as revealed in the B-footage for the broadcast that is posted on YouTube.[3] It is edited in an odd, nonlinear fashion, with long silent takes of the desolate landscape reminiscent of a college art film. In it, the soldiers struggle to get the fingerprints and images of faces to register properly. The blinking of one Afghan man and the improper pressure applied to the device by another frustrates them. One of the soldiers asks a youngish Afghan his age and is told through the translator that the young man is eighteen; the soldier scoffs and says the young man "must be twenty," entering the data into the system (it is unclear from the video which age was entered). Another shot shows the translator wearing sunglasses, a hat, and a headscarf around his face to obscure his identity.

The full video, both the news segment and the B-roll considered together, portrays the two faces of the biometric technologies developed by the U.S. Department of Defense (DoD) and deployed in Iraq and Afghanistan. One face shows efficient, reliable technologies used for capturing enemies. The other is of an unstable, contested system aimed toward the general population. The clips make it evident that the stakes are high in interactions between the U.S. soldiers and the Afghans mediated by the biometric devices. Participation poses particular risks for allies (the translator clearly does not want to be identified) and must be negotiated by both the citizens and the soldiers. Yet the results are ambiguous. It is unclear if any Taliban militants were captured or innocent citizens detained. All that is certain is that the U.S. military's biometric database is being expanded at the barrel of a gun and while surrounded by barbed wire.

Although biometrics were not used beyond limited forensic purposes before 1999, as of July 28, 2011, the military's Automated Biometric Identification System (ABIS) surpassed six million total records.[4] The shift in strategy from regime change through air strikes and Special Forces to counterinsurgency (known as "war by governance") made separating the enemy from the general population the war's primary goal and thus required the identification of individual Iraqis and Afghans. This need coincided with a shift in leadership of the U.S. military's biometric offices that saw priorities change from computer and base security to a broad application of the technology on the battlefield. These two factors helped solidify a place for biometric technologies in the face of internal opposition to doing "police work."[5] With the technology now institutionally established as the

Defense Forensics and Biometrics Agency, its proponents are working hard to maintain the program as an enduring capability that will transcend its genesis as a tool of counterinsurgency.[6]

This chapter focuses on the years when "offensive" biometrics were developed and deployed during the Iraq and Afghanistan conflicts, particularly from 2004 to 2008, to understand the nature and implications of their use.[7] It also discusses how the doctrinal debates of that time, which saw the ascendancy of counterinsurgency as the primary form of combat, shaped how the technology was promoted and used.

The complete identification of the Iraqi and Afghan populace was never completed, and biometric technologies have clear limits in actually addressing the complex social issues they are meant to definitively solve. The gap between the fantasy of a project to identify the entire population and the reality of technological limits creates an ambiguity that enables individual soldiers to make the day-to-day decisions required in a counterinsurgency campaign. Yet this very ambiguity undermines the premise on which the identification technologies rest, making every individual potentially suspect and a threat. In the end, biometric technologies become tools to objectify individuals and legitimize the strategies and prejudices of the counterinsurgents. Despite the desire to separate the population into clear categories that distinguish friend from foe, or blue forces from red forces, the population remains gray and the use of force arbitrary. While biometrics are a key component of efforts to technologically fix the identities of the occupied populace into a framework that allows clear military action, their use and limitations should instead be seen to signify the persistence of the fog of war that confounds the overwhelming power of the U.S. military and undercuts any attempt to win hearts and minds.

The observations in this chapter are based on government documents, policies, and publicity materials produced by the U.S. military and its contractors. This archival research was supplemented by on-site interviews with U.S. military personnel, contractors, and employees of the Federal Bureau of Investigation (FBI) and Department of Homeland Security (DHS) who are working on developing the military's biometric programs and by participant observation at biometric industry conferences and events. The chapter is divided into three sections. The first section charts DoD's development of biometrics from 1999 to the present, focusing on how domestic surveillance technologies played a role in their acceptance. The second discusses the influence of shifting military doctrine on the use of these technologies, and the third explores the factors that drive and impede this process.

Identity Dominance and Management: A Brief Time Line of DoD Biometrics

While the U.S. military has funded and tested biometric technologies for years, only within the last decade has a concerted effort been made to use these technologies on a mass scale and in active combat theaters. Only within the last five to six years has this project become truly operational. While the U.S. military's use of biometric technologies may seem inevitable, even in the period of rapid deployment (2004–2006) struggles surrounded the need for and the purpose of biometric databases. There appear to be three major turning points in the U.S. military's use of biometrics: the creation of the Biometrics Management Office (BMO) in 2000, the rapid expansion of BMO's scope and capabilities in response to the counterinsurgency campaigns in Iraq and Afghanistan in 2004, and the ongoing preparation by BIMA for the drawdown of forces in Iraq and Afghanistan, which requires a shift from battlefield support to the spread of biometric technologies into the daily routines of the U.S. military, writ large.

By the late 1990s, biometrics—particularly fingerprints—was over a century old and had become a normalized tool for law enforcement agencies. The U.S. military, hardly a slow adopter of new technologies, had long integrated fingerprint analysis into its investigative units but had not implemented biometric technologies beyond using them in forensic analysis and to control access to sensitive areas such as military bases or computer terminals. The military had already invested heavily in biometric technologies research through DARPA (Defense Advanced Research Projects Agency) and the Information Processing Techniques Office, but nonforensic use of the technologies (differentiated by the real-time identification aspect) had yet to be made.[8] In 2002, the DoD released a strategic plan calling for the integration and institutionalization of biometric technologies in the conduct of regular military business and battlefield operations. That same year the Biometric Automated Toolset (BAT)—consisting of a laptop computer, a fingerprint scanner, and an iris scanner—was first deployed for use by combat troops.

A turning point in the military's use of biometrics came in 2004. A suicide bombing in a mess tent in Mosul, Iraq made starkly clear how important base access had become. A memo from Paul Wolfowitz, then deputy secretary of defense, demanded biometric technology implementation to bolster defenses against this type of attack. At the same time, the BMO came under the leadership of John Woodward Jr., a Rand scholar and strong proponent of biometric technologies. Under Woodward's tenure, the BMO expanded the depth and scope of its work immensely, specifically changing its focus from identity verification of U.S. military personnel to the tracking and identification of enemy forces. In 2004, a year after the United States and its allies invaded Iraq, the DoD established the ABIS, its current biometric database. Woodward recalled stiff opposition to the expansion of biometrics onto the battlefield, with critics claiming such "police work" wasted precious resources.[9] Through the support of sympathetic lawmakers and the FBI, this dissent was overruled, thereby paving the way for the use of biometrics as an "offensive capability." The DoD also cites 2004 as the first successful match in the field, in which a detainee's fingerprints were matched to an individual who had previously been captured in 2003.[10]

In 2005, Woodward published an article in *Military Review* calling for "identity dominance," by which the military can quickly distinguish "friend from foe" by linking an individual to his or her previous identities and possible criminal or terrorist histories.[11] That year the U.S. military established the Biometric Identification System for Access (BISA), which encodes biometric traits on smart cards for access to military facilities by coalition forces, allies, and employees. At this point multiple active biometric databases were used in Iraq—ABIS, BISA, the limited (but regularly updated) databases deployed with BAT, and the Iraqi Automated Fingerprint Identification System (AFIS), which was built on Saddam Hussein's criminal files and enabled the Iraqi government to screen potential employees.[12]

A second tactical system, the Handheld Interagency Identity Detection Equipment (HIIDE), was developed and deployed, boasting greater portability (it looks much like a large camera) but a more limited database. In April 2006, the BMO and Biometrics Fusion Center were combined into a single entity, the Biometrics Task Force under the command of the army's chief of operations. In 2007, during the battle in Fallujah between coalition and insurgent forces, the U.S. Marines walled off the city and closed its roads, allowing movement of only those who had received a "badge"—granted after biometric registry with BAT.[13] Fallujah being an active war zone meant that most noncombatants were told to evacuate the city (to be registered on reentry). Those who remained were viewed as combatants to be targeted.[14] The perceived success of biometrics use in Fallujah helped bolster the case for technological expansion across departments and contexts.

Biometric enrollments were also collected as part of employment applications, checkpoint operations, and explicit information-gathering missions. Prisons were one of the more important sources of biometric data, given the high release rate of inmates. Bing West and Owen West claim in a 2007 *New York Times* op-ed that, "according to Pentagon records, more than 85 percent of the suspected Sunni insurgents and Shiite militiamen detained are soon set free. . . . The troops call it 'catch and release.'"[15]

Whereas the Wests see detainee releases as "vastly excessive civil rights protections for detainees" brought on by the "shameful abuses of Abu Ghraib," the term "catch and release" compares human detainees to hunted animals. Thomas Barnett, a professor at the Naval War College who advised the Office of the Secretary of Defense, explicitly stated that the military engages in "passive collection" of Iraqi biometrics "because these guys will scatter over time. . . . [W]hen you have the opportunity to tag them, you tag them before you release them into the wild."[16] In other words, mass arrests and neighborhood sweeps that resulted in few actual convictions were quite productive. By the end of 2007, ABIS contained over 1.5 million entries.

In May 2010, the secretary of the Army signed Department of the Army General Orders that redesignated and transformed the Biometrics Task Force into BIMA, making it a permanent organization with full funding and staff. Its records are technologically compatible with those of the FBI and the DHS.[17] It has also been reported that as of October 2008 at least seven thousand DNA samples were taken from Iraqi and Afghan detainees and deposited into the Joint

Federal Agencies Antiterrorism DNA Database, with an additional ten thousand "inbound."[18]

The DoD has contributed significant funding to the mass production of BAT (and more recently, handheld devices such as the HIIDE) units, driving down prices and spurring rapid innovation focused on portability and durability. These technologies circulate through trade shows, academic conferences, official working groups (such as the Biometrics Consortium), and personnel exchanges among the various organizations in a manner that blurs the border between technological systems for occupation and for domestic security.

The flow of these technologies is bolstered not only by economies of scale and presidential directives ordering cooperation but also by a larger merger of techniques, technologies, and goals between the agencies tasked with security and dealing with national "others" (i.e., criminals, immigrants) and the subjects of military occupation. While many have already quite convincingly argued that domestic policing and border control have been militarized,[19] I argue that the military has seen the "policification" of its duties, objectives, and tactics. In other words, policing and warfare are converging. It is a state of affairs easily recognizable in the organizational liminality of the war on drugs and the war on terror. The tension between the functions of policing and war fighting, and the growing speciousness of the dichotomy it represents, was prominent in the 2004 U.S. presidential campaign when Senator John Kerry argued that terrorism is a matter of "law enforcement," a position that President George W. Bush's campaign derided as a "pre-9/11 view of the world" that did not "understand the threat" and was not properly focused on "winning the war."[20] Yet these differences were mainly semantic, because in the second term of the Bush administration the military in Iraq took on the very tools of law enforcement, such as regular foot patrols (walking a beat), mediating local disturbances, and tracking suspicious individuals to build evidence for criminal prosecution (through forensics and biometrics).[21]

Competing Doctrines

The offensive use of biometrics by the U.S. military has occurred against the backdrop of two active counterinsurgency campaigns, in

Iraq and Afghanistan, and had its genesis in smaller-scale operations in Kosovo. Their development and deployment have been heavily influenced by a broader context of competing military doctrines and political struggles within the Pentagon.[22] The internal conflict of the past decade can be described as the clash between two discordant visions of the future of warfare. The first vision is exemplified by the revolution in military affairs (RMA) as argued for by Admiral William Owens and at times referred to as the Rumsfeld doctrine after former secretary of defense Donald Rumsfeld.[23] RMA is characterized by an emphasis on developing and deploying networked communication and satellite technologies, along with small, mobile units of troops and a heavy reliance on air strikes. The military envisioned by RMA would dominate the battlefield through insurmountable technological superiority while minimizing risk for its own troops. RMA is exemplified by targeted Predator drone strikes, based on satellite imagery and reports from clandestine intelligence units.

The second is articulated by the counterinsurgency (COIN) doctrine as outlined in the *Counterinsurgency Field Manual* and championed by General David Petraeus.[24] Under COIN doctrine, the military relies heavily on large troop deployments with deep penetration of the occupied population and an emphasis on activities more akin to policing and governance than outright warfare. Although an old staple of warfare, most modern counterinsurgency theory stems from reflections on the battles that were fought throughout the Cold War and struggles for colonial independence. Its revival was made famous during the surge of U.S. troops in Iraq in 2007 to quell the growing insurgency. A similar strategy was aggressively deployed in southern Afghanistan in 2012.

There are readily apparent and subtler similarities and differences between these doctrines. Both stand in direct opposition to the long-dominant Cold War doctrine of deterrence through dominance of conventional capacity (measured in terms of numbers and sophistication of nuclear stockpiles, aircraft carriers, and tank battalions). While RMA tilts toward conventional warfare, both it and COIN assume that warfare of the future will be asymmetrical. In other words, the United States will be engaged in conflicts with enemies possessing vastly inferior conventional capabilities. Both principles envision an enemy who wears no uniform, blends into the population, and strikes from the shadows. Thus, intelligence gathering and the ability to clearly identify the hidden enemy is key to both. Keith L. Shimko makes this similarity clear by emphasizing the importance of the term "situational awareness" to both RMA and COIN proponents. Whether this enemy is primarily engaged through surgical strikes or sustained contact with the local population has been the subject of fierce debate within the Pentagon over the last ten years, with the consensus during the time of increased use of biometrics by the U.S. military heavily favoring COIN doctrine.[25]

The use of biometrics has subtly grown and shifted with doctrinal changes. While initially a strong component of the future tech that would provide informational dominance over the enemy (for example, through facial recognition technology attached to drones), biometrics is becoming a more mundane layer of security for entry to bases and the management of local governance. This is not to say that biometrics does not still hold a degree of technological shock and awe intended to dissuade resistance but that, like the rest of the technological fantasies of the RMA founders, the limitations of actual infrastructure and the shifting realities on the ground have mutated and altered perceptions about the potential for biometrics. This can be seen as playing out in the shifting terminology from "identity dominance" to "identity management," in which the goal of hunting down bad guys is tempered by the greater need to identify the soldiers, nongovernmental organization personnel, and civilians who need access to bases, supplies, and assistance.

It is useful to look a bit closer at the COIN paradigm to understand the burgeoning role of biometrics in the U.S. military.[26] In 1965, the French counterinsurgency theorist Bernard Fall wrote that "a government that is losing to an insurgency isn't being outfought, it's being outgoverned."[27] David Kilcullen, counterinsurgency expert and an advisor to General Petraeus, put this in even starker terms, stating that counterinsurgency is essentially "armed civil affairs" or "armed social work."[28] Put simply, the counterinsurgent's goal is to create or maintain the basics of state power. The counterinsurgent seeks to ensure the security of the populace, provide civil services (such as mediation of disputes and dispensation of justice), and guarantee the material and social welfare of the populace. The goal of the insurgent is essentially the same, but first the insurgent must sufficiently destabilize the counterinsurgent's ability to maintain order before filling the power vacuum. Because the insurgency-counterinsurgency tension is a battle of governance, all institutions of governance are contested, often producing rival aid agencies, court systems, and even marriage ceremonies. The military during a counterinsurgency campaign is thus transformed into a ministate, focused on policing, the state of local politics, and the meeting of civic and social needs.

The need for identification in counterinsurgency is therefore quite urgent. David Galula, another mid-twentieth century French counterinsurgency theorist, argues, "Control of the population begins obviously with a thorough census. . . . Every inhabitant must be registered and given a foolproof identity card."[29] This serves the primary purpose of gaining basic intelligence and gives the counterinsurgent tools to begin separating insurgents from the population, or to paraphrase Mao Tse-tung, the "fish from the water." More importantly, however, a census is one of the fundamental means of establishing the structures of modern, bureaucratic governance. As John Torpey argues, "The reach of the state . . . , cannot exceed its grasp," or in other words, the state cannot govern (for better or worse) individuals with whom it has no relationship.[30] The most rudimentary of means of the state to "embrace" its citizens is to take a census,[31] although the historical variations of what a census collects and tabulates indicates both the state's grasp and its priorities.[32] That identification is a form of contested governance during an insurgency or counterinsurgency is made evident in the Taliban producing and distributing their own identity cards, as discovered by the U.S. military in an abandoned base of operations in Afghanistan.[33]

Censuses have been the primary means through which states make their citizens visible and, in the words of James C. Scott, "legible" and are by extension a staple of counterinsurgent warfare. Censuses used by occupation forces can be traced at least as far back as the Roman Empire and were used extensively in the second half of the Vietnam War under the auspices of Operation Phoenix, the ill-fated counterinsurgency effort.[34] Censuses by the U.S. military under COIN are perhaps the most sophisticated efforts yet deployed, using sociologists and linguists in human terrain teams, not without controversy.[35] The history of the use of identity cards in armed conflict, as suggested by Galula, is somewhat murkier and is most certainly a modern phenomenon.[36] A growing body of literature addresses the rise and role of identification cards[37] and even their use in scenarios of occupation and conflict.[38] As argued by Helga Tawil-Souri in discussing the use of ID cards by Israel to maintain surveillant control over Palestinians, "ID cards serve as a point of physical and tangible contact between Palestinians and the state of Israel . . . the [ID] card becomes the physical substance through which their relationship is mediated."[39] In spaces of conflict, just as in normalized states of bureaucratic governance, identification cards are a necessary material substantiation of the relationship between an individual and the state and are one of the primary means through which one can relate to or place demands on the other. Under conditions of insurgency, these relationships are strained, contested, or even nonexistent, and it is incumbent on the counterinsurgent to (re)establish the media through which it can embrace the individuals under its purview. The lack of an adequate and immediate census following the invasion of Iraq, with the enforced dispersal of identification cards, was one of the sharper criticisms levied by the disciples of COIN against Rumsfeld's occupational tactics.[40]

The drive to separate the enemy from the populace is not unique to COIN. Just-war theory, which has long theoretical roots, is largely predicated on the protection of noncombatants, who must somehow be identified as such. The role of biometrics in RMA (e.g., facial recognition technology on drones) is indicative of the need for some degree of population segregation even in the realm of violence. As argued by Michael Dillon and Julian Reid, war pursued by liberal governments reflects their biopolitical drives.[41] It is "killing to make life live" or, to be more specific, killing as a part of the creation and protection of a particular form of "biohumanity." However, the categorization of individuals between those who may live and those who must die in a counterinsurgency is not merely a means to an end but rather the end in itself. It is the central strategic goal. The scale of identification and the urgency required in warfare are vastly greater than the mere creation of watch lists or forensics. Counterinsurgency requires the registration and tracking of all those who encounter the military across time and context. Ideally, it would include the entire population.

The technological capabilities of biometrics are particularly useful in achieving this task. Biometrics are a black box technology that creates legibility of individuals through traits that are not legible to themselves and thus not easily mutable. More importantly, biometrics (as opposed to an ID card) create a database that can be centrally stored, networked, and infinitely reproduced. This enables knowledge gained by individual soldiers to collect and pass on knowledge that can be analyzed and used both by the central command and by their compatriots in other locations long after the initial soldiers' rotations are finished. As knowledge of the local populace is key to the creation of the ideal counterinsurgency, this centralized and replicable knowledge base helps (theoretically) ease the transition between troop deployments. Finally, the forensic capabilities of fingerprints became a key tool in the fight against improvised explosive device (IED) attacks, leading Marine Corps Lieutenant Colonel Thomas Pratt to call biometrics a "killer app."[42]

In short, counterinsurgency is war conducted through the tools of modern governance, and modern governance requires the identification of individual citizens because identification is the base unit of legibility of the modern nation-state. This connection helps explain the introduction of tools of domestic surveillance and control into the war zones of Iraq and Afghanistan. A general convergence of goals and tactics between the military and the police has been a necessary component of this dynamic, which has included a greater degree of technological circulation rather than a one-way transfer from foreign engagements to domestic control—this is particularly visible in the development of battlefield biometrics. Importantly for the recent counterinsurgency campaigns, biometric technologies created an opportunity to centralize the knowledge captured through interactions with the local populace. It could then be used as a substitute for the local understanding needed by soldiers on the ground, which is otherwise routinely lost with each troop redeployment. It is important, however, to look at how the promise of biometric technologies that helped drive their rapid deployment in Iraq and Afghanistan performed and what that meant for their actual use and utility.

Red, Blue, and Gray

Biometric data are only as useful as the information attached to them. A fingerprint without context is nothing but a smudge. This is not to say that there have not been attempts to read biographical data into biometric signifiers. Even fingerprints were thought at the turn of the twentieth century to yield what Francis Galton dubbed the "biological coat of arms,"[43] and today studies are still being produced that link sexual preferences with fingerprint patterns. Other biometric traits have an even greater potential for storing biographical information. Faces and irises can be construed to enable racial classification, and several companies are built on the premise that a DNA sample can provide hidden genealogical, health, or psychological data.[44]

Whether biometric traits themselves enable simple verification or identification projects or are harbingers for profiling has been a point of often-contentious debate within the biometrics community. Simon Cole has thoroughly documented the (very successful) efforts of early fingerprinting experts to marginalize those who would read biographies in the loops and whorls.[45] Similarly, profiling technology and processes often take pains to remove the individual from the equation through anonymization and aggregation. If a characteristic or trait is truly unique to the individual, then it is useless to the profile, as it does not allow for categorization or prediction of other individuals. Put simply, identification looks for unique traits among the masses, whereas profiling searches for the mass traits within the individual.

This distinction is vitally important for the privacy implications of identification technologies because it establishes some degree of what Helen Nissenbaum calls "contextual integrity." Nissenbaum states that the dichotomy of "public" and "private" is erroneous and counterproductive and that in "observing the texture of people's lives, we find them not only crossing dichotomies, but moving about, into, and out of a plurality of distinct realms."[46] The concern that people have when discussing harms to their privacy is the proper flow of information within and between these contexts. For example, you would not share the same kinds of information with your doctor as you would with your employer, even though you can be reasonably required to share information with both that could be considered private. As Nissenbaum states, "What most people care most about is not simply *restricting* the flow of information but ensuring that it flows *appropriately*."[47] Universal identification programs, and biometrics in particular, if not properly configured, carry with them the potential to destroy the boundaries between various contexts. Information that is attached to an identity in one context can be accessed improperly in another through the common system. Yet these violations can occur only if biographical data is attached to identifying data that can be cross-referenced over time and by various organizations or agencies. This may seem mundane in a war context, a First World problem, but it is in fact quite relevant. When the criminal files created by Saddam Hussein are used as the basis for Iraqi AFIS or when information collected on a neighborhood sweep is used to deny asylum status, the barriers between context of collection and context of use can be matters of life or death.

The U.S. military is aware of privacy concerns and the most extreme ramifications of biometrics' misuse. "Report of the Defense Science Board Task Force on Defense Biometrics" outlines various methods of designing technologies such that they pose a "lower" or "greater" privacy risk—for example, whether a subject knows he or she is being tracked. Despite the acknowledgment that biometrics can be designed to not harm but in fact protect privacy, the authors conclude, "Sadly, but not unexpectedly, for most DoD applications of biometrics in support of identity management, the more desirable characteristics . . . involve a greater risk of privacy invasiveness."[48] Lieutenant Colonel John W. Velliquette Jr. admitted that "[ABIS] is also very sensitive, because essentially what it becomes is a hit list if it gets in the wrong hands."[49] Information collected on U.S. soldiers is covered by the protections of the Privacy Act of 1974, and while restrictions are in place on what type of data can be collected from coalition partners and what can be done with it, no such limitations were placed on the information gathered from Iraqis and Afghans.[50] The information that DoD attaches to the biometric file of an individual is as comprehensive as possible. According to a military biometrics tactical manual, these are the categories of information that would be attached to an individual's biometric:

(1) Biographic.
  (a) Name.
  (b) ID number (e.g., passport, national ID, detainee number etc.).
  (c) Phone number.

(d) Date of birth.
(e) Place of birth.
(f) Occupation.
(g) Gender.
(h) Place of residence.
(i) Family members' names.
(j) Marital status.
(k) Organization memberships.
(l) Other names used.

(2) Contextual.
(a) Location of encounter (e.g., military grade reference system, grid location, or description of enrollment site).
(b) Reason for enrollment (i.e., person of interest, census, base access, detainee operations).
(c) Date and time enrolled.[51]

Almost all of these entry points have been incorporated into the biometric enrollment systems with several open sections for more information. The BAT system also records date and location of each encounter with an individual.[52] As evidenced here, empty identifiers—fingerprints, irises, and pictures—are quickly filled with biographical and geographical data. Importantly, relational data is also included, such as tribal affiliation and the names and Civil Affairs numbers of all immediate family members. It was this relational information that Lieutenant Colonel Velliquette was concerned could be turned into a "hit list," because political, religious, or ethnic affiliations are often inferred through family names and tribal membership.[53]

Given these risks, why does the U.S. military collect so much information? This question becomes even more acute when seen through the lens of the limited military budget: the commander on the ground must decide between buying a BAT system or more M16s. Even the truncated description of the DoD's biometrics programs in this chapter is distilled further by the promotional materials, government reports, and industry materials into a simple, underlying directive: finding the bad guys (or its corollary of separating the good guys from the bad guys). While this description may seem an oversimplified caricature of a militaristic, Manichaean worldview, it is surprising how often the terms "good guy" and "bad guy" come up in DoD materials and in conversations with individuals working on these projects.[54] A fingerprint, however, does not grant this type of insight into a person's essential character. Biometric data does, however, enable the connection of disparate events, locations, and categories (e.g., a latent fingerprint at an IED site, a tribal identity thought to be hostile, an age range) such that a profile of risk and possibility of harm, past and future, can be created and maintained. These profiles of risky identities create the visibility necessary for kinetic action. They produce the targets of military action.

The technical terms for "good guys" and "bad guys" are, respectively, "blue forces" and "red forces," with blue being clear allies and red being clear enemies. The importance of blue force biometrics has risen significantly since the inception of the DoD biometrics program, as indicated by the transition in terminology from "identity dominance" to "identity management."[55] In "identity dominance," the focus is on identifying, tracking, and capturing enemy forces while denying the same enemy the ability to do the same to U.S. assets. "Identity management," on the other hand, recognizes the need to both grant access and detain, to sustain life and mete out death. It is an explicit recognition of the biopolitical potential of biometric technologies—that biometrics can be used just as much to manage populations as it can to identify the enemy. The shift from RMA's enemy-centric strategies of precision strikes and remote war toward the population-centric approach of COIN during the period of biometric expansion and institutionalization also helps explain the rise of the term "identity management," wherein the day-to-day business of governance is of a higher priority than militarily dominating an asymmetric foe. The use of "identity dominance" is beginning to make a comeback, being described prominently in BIMA's 2011 annual report and 2012 promotional video. This can be seen as a subtle shifting of discourse now that the counterinsurgency campaigns are being phased out and sole reliance on airstrikes and Special Forces is becoming (again) more common.[56]

This discursive shift is particularly evident in the inclusion of the term "gray forces" in some military documents describing the mission of BIMA and its related agencies. "Gray forces" describes individuals or entities whose relationship or motivations are "not clear."[57] This could include nongovernmental organizations, all local nationals, foreign politicians, third-country nationals, and local employees. U.S. Central Command is quite clear as well that the "majority of [the entities its personnel encounter] are gray."[58] Indeed, the situation quickly becomes less clear. For instance, are sources such as Ahmed Chalabi (whose loyalties shifted repeatedly throughout the Iraq war) or the suicide bomber who gained access to a Central Intelligence Agency outpost after being actively cultivated by the same people he killed red, gray, or blue? Is Ahmed Wali Karzai, brother to an Afghan president, who was before his death a regular ally of coalition forces *and* a noted drug dealer, red because his illicit activities dominated his other connections and history? In short, even while the military recognizes that gray is the predominant color of the population, even its designations of red and blue are uncertain given that individuals can have multiple identities.

This reiterates the importance of visibility to recent U.S. military doctrines. The U.S. military is confident that in conventional warfare its capabilities remain unrivaled. It currently accounts for 46 percent of worldwide military expenditures, which is seven times more than its closest rival, China.[59] Yet the United States is faced with threats, both active and imagined, that are asymmetrical, unconventional, and coming from enemies who are hidden, dispersed, and embedded within the communities the military is charged with protecting (both domestic and international). The rhetoric that pervades discussions of biometric technologies is all about being able to tell the good guys from the bad guys, yet in truth, the promise of biometrics is to summon forth the blue and red forces from the unknown mass of gray.

Ideally, all individuals within a population are easily categorized as "friendly" or "enemy," but this is impossible for many reasons. First, in a counterinsurgency the population as a whole is neither friendly nor hostile; it is their hearts and minds over which the battles are fought. Battles are fought less over territory than over legitimacy, with insurgents attempting to open vacuums to fill and the counterinsurgents attempting to maintain hegemony. The loyalty of the population is more than simply in play; it is the prize. Thus, red can switch to blue and then switch back depending on the tides of war. Amid this fluctuation, the use of biometrics is an intentional and active attempt by the U.S. military to fix the identities of individuals as friendly. To conduct daily business and to interact normally with

Coalition forces, an individual must be registered, and this registration is promoted as an act of tacit support. Galula argues for placing responsibility on the population itself for verifying and updating the information gathered in a census, in turn forcing them to "participate willy-nilly in the struggle . . . [and] . . . contribute to turn[ing] the population against the insurgent."[60] This means that those who may otherwise be ambivalent about the conflict are forced to actively participate in governance activities and, by extension, in the counterinsurgency. This places the insurgency in a difficult position—active participation in and support of the counterinsurgent governance project cannot be tolerated, but too harsh a response to coerced or negligible compliance will turn popular opinion against the insurgency. Loyalties and motivations change, and counterinsurgents rely on this. Unlike the explicit indiscriminate violence promoted by some in the early stages of the war on terror, counterinsurgency theorists argue that a war is over not when all hostile forces are dead but rather when they choose to put down their arms.

The vast majority of the population remains gray because people's motivations are too complex to fit into a Manichaean binary. To use Scott's terminology, it is necessary for the rationalized purposes of the state to transform the "social hieroglyph [i.e., the complexity of reality] into a legible and administratively more convenient format."[61] This occurs through the "rounding off" of difference and ignoring divergences from the norm to maintain the integrity of the system. Yet the state is well aware of its rounding errors, as is shown in the recognition that "the majority of [people] are gray."[62] Biometric technologies are a response to the gray fog of an unknown, complex population.

As a solution, however, biometrics are fraught with their own set of problems. First, biometrics are vulnerable, as are all identification technologies, to enrollment fraud—to individuals falsely representing themselves when initially entered into a system. An individual entered under a false identity will be able to pass as that individual upon all future identity checks and in fact has his or her false identity legitimized by enrollment in the system.

Second, operators of biometric technologies also limit the ultimate capabilities of those technologies. In June 2009 Myra Gray, then director of BIMA, implored U.S. troops to make sure that "every fingerprint, every iris scan is . . . done correctly and precisely" or else "the potential of exposing our troops, as well as our families at home, is greater . . . our screenings to detect terrorists are only as good as the databases we build."[63] The need for, and more importantly the lamentable inadequacy of, local clerks and sergeants to carry out the duties of attention are a driving factor in advancing surveillance technologies since the invention of the modern passport system in revolutionary France to state-of-the-art biometric processing algorithms employed today.[64] This automation of surveillance technologies is part of an ongoing process of deskilling labor, particularly that of the police. In this way, the use of biometrics in Iraq and Afghanistan can be seen as an attempt to make up for the inability of ground commanders to live up to the almost ridiculous demands of COIN, especially given the finitude of any given commander's deployment in a particular region.[65] Yet even in centralizing and standardizing the knowledge built from each subsequent deployment, commanders depend on the reliability of the data collected by all those who proceeded them, often with mixed results.[66]

Biometric technologies as a form of security are also fundamentally vulnerable to what former Homeland Security director Michael Chertoff called "clean skins"—individuals with no recorded history of "criminal" acts or associations.[67] This rhetoric is quite evocative; it entails a "dirty" body underneath that needs to be discovered and further implies that there exist individuals who have "dirty skins" but "clean bodies"—innocent individuals with records that have been tainted by association with violent acts. This problem causes biometrics (and indeed, all forms of idem identification) to fall silent.

Third, the very development of biometrics by the DoD was itself not without problems. Cross-agency sharing of data, even within DoD, is a challenge technically as well as institutionally. In fact, while it appears that many of the technical hurdles are finally being overcome (it was just in 2011 that ABIS became compatible with DHS files), bureaucratic power struggles over ownership of data as well as differing regulations over the sharing and use of individuals' information continue to present obstacles.[68] In 2010, the National Research Council published what was characterized as a "withering critique" of biometric technologies, stating that "in spite of substantial effort . . . there remain unresolved questions about the effectiveness and management of systems for biometric recognition and societal impact of their use."[69] In particular, the report pointed to a need to recognize the "probabilistic nature" of biometric technologies and that decisions based on their use should be tempered with "an awareness of the uncertainty associated with the recognition."[70] It also raised the concern that no biometric trait has been proved to be "extremely stable and distinctive across all groups."[71] This point is exacerbated by the fact that it is known that some traits are particularly indistinctive among specific groups. For example, fingerprints are hard to collect from manual laborers and are not stable in children under age five. While it has been suggested that multimodal biometrics will help limit the problems caused by lack of a single modality, one study found that multimodal systems are just as, if not more, vulnerable to spoofing as single-mode biometrics.[72]

In a BIMA promotional video that otherwise extols the virtue of the futuristic technologies that help the "warfighter" distinguish "friend from foe," Sergeant John Stroud states that "[biometrics] is similar to a home security system; most people have the signs out in their yard. They may not have [the security system], but it does keep the enemy away."[73] To be clear, Stroud is not comparing biometrics to the home security system; he is comparing it to the *sign*. In other words, the power of biometrics lies less in its actual capabilities and more in its perceived capabilities. This surprising admission by Stroud echoes William Bogard's description of the importance of the "imaginary of surveillant control," wherein the perception of a totality of surveillance influences policies and development and, in the words of Kelly Gates, "functions as a form of social control" regardless of actual capabilities.[74] While Gates makes a good point that the importance of the "imaginary of surveillant control" should not preclude a study of the development of particular technologies and designs as sites of indeterminate political struggle, there is nonetheless an important interplay between the constructed infallibility of surveillant assemblages and their actual (in)capabilities.

The fantasy of biometrics, similar to Michel Foucault's description of the quarantined town in which each individual was assigned "his 'true' name, his 'true' place, his 'true' body, his 'true' disease,"[75] was simulated in Fallujah and in other locations in Iraq and Afghanistan. Yet this level of control is fleeting without immense expenditure of resources, and even then many of the technical issues discussed above preclude its sustained (if ever initially achieved) totality. As stated by Stephen Graham:

> Technological "silver bullets" . . . often fall far from their target. They fail to function, continually break down, do not deliver the anticipated results, and do nothing to address the root causes of feelings of insecurity. . . . [T]he techno-dreams fail simply because the technology breaks down or fails to mesh with a myriad of other technologies or because operators are unable to deal with the system's complexity.[76]

Yet while Graham argues that this bars a "global Panopticon," it is precisely the Panopticon that provides the metaphor with which to understand the gap between technological fantasy and reality. The Panopticon, as described by Foucault, is structured such that there is only one guard present or not present at all.[77] The prisoner may be watched at all times *or not watched at all*. The Panopticon is a totalizing system of surveillance only inasmuch as the prisoner "assumes responsibility" for its control.[78] It is a system of power with minimal resources, a sign that may or may not have an alarm behind it.

This has a profound influence on the ultimate impacts of biometric technologies. As stated earlier, biometrics were introduced to centralize and stabilize knowledge about individuals encountered by the U.S. military for use by soldiers who are expected by the dictates of counterinsurgency to maintain a high degree of contextualized information despite rotations, limited deployments, and human shortcomings. Yet the relationship between the soldier and the populace is necessarily murky. It is advantageous to insurgents to blend as much as possible into the populace and to exploit patterns or behaviors that would enable them to do so. Soldiers are trained to follow particular rules of engagement with the populace and to use technologies like biometrics to help segregate the population at checkpoints, but they must exercise discretion at each turn, looking for the insurgent who does not register. As stated by a former checkpoint solider with the Israeli Defense Force (IDF):

> The soldier's responsibility to interpret any given case as an exception to the rule is part of the IDF's general strategy to undermine its own patterns and regularities. The army doesn't want Palestinians to be able to foresee what might get them through the checkpoint quickly and safely. The clear-headed judgment clause indirectly prevents exploitable patterns of behavior from emerging.[79]

The military deploys tools of identification to clearly reify and demarcate the population it is engaging so as to summon forth its enemies. Yet how these demarcations are created must remain shrouded and fluid. Thus, even those individuals sympathetic to the goals of the counterinsurgents cannot signal their allegiance. With or without biometric technologies, each Iraqi is a possible suicide bomber, each Afghan potentially a Taliban member.

Biometrics is a part of this destabilization of exploitable patterns. While ID cards are at least partially legible to the holders—they are able to check to see if their name, photo, or address matches that on the card—a fingerprint or iris does not reveal its secrets without advanced algorithms or technical experts. In addition to being practically unable to separate themselves from their biometric traits, individuals are also generally ignorant of what information they contain. This means that each encounter with the U.S. military, regardless of guilt or previous interactions, is tinged with uncertainty and suspicion. A blemished record may guarantee a negative response from American soldiers, but a positive or spotless record does not protect one from harm. To move from gray to red is easy, whereas moving from gray to blue is practically impossible. This plays itself out each time a soldier must decide to let a citizen pass or to detain her or him, and the introduction of biometric technology merely adds an aura of technological legitimization to the interaction.

Conclusion: The Persistent Fog

The U.S. military is reassessing its biometric programs with a focus on slowly shifting back from red force identification to blue force identity management. This is happening as the pendulum of asymmetric warfare has shifted back toward proponents of air power and limited ground forces, using the war in Libya that ousted Muammar Gaddafi as a "new" paradigm. Having been established and institutionalized, the red force programs are hardly disappearing, and integration with the databases of the DHS and the new multimodal database of the FBI continues its slow but steady pace. Even in the Special Forces–type operations integral to RMA-type attacks, such as the Navy Seal team execution of Osama Bin Laden, the technological wizardry of facial recognition technology was touted as having played a crucial role. The military's offensive use of biometrics, while still in its infancy, is likely to play a major role in future conflicts, despite domestic budgetary concerns.

The necessity of biometric technology, especially given its current actual capabilities, is less an expression of absolute technical dominance and more of recognition by the U.S. military, the possessor of the largest destructive power in history, of its own blindness and limitations. Biometric technologies also signify the intrusion of care for life into the realms of violence, and while these strategies are two sides of the same governmentality coin, they nonetheless sit uneasily with one another. This conflict plays out in and through the uses and development of technologies such as biometrics, with the focus being placed at times on watch lists and mobility, at other times on employee management and centralized control. Regardless of application, the gap between the fantasies of a clearly demarcated population revealed through technological omniscience and the persistent gray of the actual populace allows for the legitimization of particular tactical decisions while denying clear routes for locals to remove suspicion and gun muzzles from themselves. Rather than drawing forth a true picture of the occupied population, biometric technologies reproduce the heavy-handed abstractions and simplifications of the occupiers.

The use of biometrics in Iraq and Afghanistan should not be seen as an overcoding of the identities of these populations or as akin to the use of that tattoo in Auschwitz. Instead, it should be understood as an attempt to create a materialized link between the U.S. military and the populations of the occupied countries through which life could be maintained and managed (in part, through the death of those deemed a threat to the particular political configuration of life worth protecting). This project, like the counterinsurgency itself, was a governmental project that was constantly undermined by the governments being supported through its efforts, as no amount of technical expertise or localized knowledge can overcome the legitimacy crisis caused by corrupt or incompetent local leadership. While it is touted as a success and an enduring capability worthy of continued financial support, this is only because of its marginal successes in capturing a (relatively) small number of insurgents, most of whom had already been captured at one point or another. Instead, the work that the biometrics program did in practice was to help legitimize who should be kept, who should be let go, and who should be shot on the spot for the troops on the ground, while keeping the outcome of these decisions opaque and unknowable to the populace.

This dynamic is unlikely to change anytime soon, with the dueling paradigms of counterinsurgent occupation and RMA-style tactical strikes being two sides of the asymmetrical war coin. At least for the foreseeable future, this form of warfare will continue to be what drives international conflicts. As long as the populace is both the object of protection and the source of potential hazard, technological developments will be relied on to answer the difficult, changing, and in many ways unknowable question of who is an enemy and who is a friend. In a world of porous borders, this concern has fused with surveillance of the domestic populace. Biometric technology and its advances are just one piece of this assemblage, yet it is a part that is particularly opaque and indelible to those it touches.

Notes

1. Quoted in Laura Blumenfeld, "Spurred by Gratitude, 'Bomb Lady' Develops Better Weapons for U.S.," *Washington Post*, December 1, 2007, available at http://www.washingtonpost.com/wp-dyn/content/article/2007/11/30/AR2007113002302.html.
2. American Forces Network Afghanistan, "Soldiers Collect Biometric Data during Counterinsurgency Operations in Khost Province," *YouTube*, November 9, 2010, available at https://www.youtube.com/watch?v=XADBQCrUBT8.
3. American Forces Network Afghanistan, "There's Nowhere to HIIDE (Hide) in Khost Province," *YouTube*, November 9, 2010, available at https://www.youtube.com/watch?v=yEexaPFVLRo.
4. Biometrics Identity Management Agency, *Annual Report, FY11* (Washington, DC: Biometrics Identity Management Agency, 2012), available at https://web.archive.org/web/20120621051935/http://www.biometrics.dod.mil/Files/Documents/AnnualReports/fy11.pdf.

5. John D. Woodward Jr., interview by the author, Washington, DC, April 27, 2011.

6. The agency's website is available at http://www.biometrics.dod.mil/.

7. Biometrics Task Force, *Annual Report, FY08* (Washington, DC: Biometrics Task Force, 2009), 44, available at https://web.archive.org/web/20110904220216/http://www.biometrics.dod.mil/Files/Documents/AnnualReports/fy08.pdf.

8. In the United States, the FBI has historically been at the forefront of funding for biometric technologies, but in the mid-1960s and at the height of the Cold War the research arms of the military became increasingly invested in the technologies as well. See Kelly Gates, *Our Biometric Future: Facial Recognition Technology and the Culture of Surveillance* (New York: New York University Press, 2011); Arthur L. Norberg, Judy E. O'Neill, and Kerry J. Freedman, *Transforming Computer Technology: Information Processing for the Pentagon, 1962–1986* (Baltimore, MD: Johns Hopkins University Press, 1996); John D. Woodward Jr., Nicholas M. Orlans, and Peter T. Higgins, *Biometrics* (New York: McGraw-Hill/Osborne, 2003); and National Science and Technology Council, "Biometrics History," August 7, 2006, available at http://www.biometrics.gov/documents/biohistory.pdf.

9. Woodward, interview by the author.

10. Donna Miles, "Biometrics Helping Identify Foes in War on Terror," November 5, 2004, available at http://archive.defense.gov/news/newsarticle.aspx?id=24912.

11. John D. Woodward Jr., "Using Biometrics to Achieve Identity Dominance in the Global War on Terror," *Military Review*, September–October 2005, pp. 30–34. "Identity dominance" is a particularly interesting phrase, for it implies that the United States not only would be able to deny its adversaries anonymity but would also deny them access to the tools of identification. I was told by Min Chong, a former member of the BMO and current CEO of Strategic Operating Solutions Inc.—a defense consulting firm specializing in biometric systems—that this aspect of "identity dominance" is classified information but that it is a factor of the DoD's biometric projects. In other words, the military is actively adopting measures to confound efforts at identification of particular military personnel or assets. It is possible to see how this would be important for espionage or other intelligence gathering activities. Yet that the U.S. military is actively undermining shared systems of identification is an ironic, if predictable, twist. Min Chong, interview by the author, Arlington, Virginia, December 1, 2010.

12. John W. Velliquette Jr., "The Role of Biometrics in the Counterinsurgency," *Department of Defense Bloggers Roundtable*, August 15, 2007.

13. Noah Shachtman, "Iraq Diary: Fallujah's Biometric Gates," *Wired*, August 31, 2007, available at http://www.wired.com/2007/08/fallujah-pics.

14. Benjamin Müller, *Security, Risk and the Biometric State: Governing Borders and Bodies* (London: Routledge, 2010), 111–115.

15. Owen West and Bing West, "The Laptop Is Mightier than the Sword," *New York Times*, June 15, 2007, available at http://www.nytimes.com/2007/06/15/opinion/15west.html.

16. Quoted in Jim Krane, "U.S. Military Compiles Biometric Database on Iraqi Fighters, Saddam Loyalists," *Information Week*, March 4, 2003, available at http://www.informationweek.com/us-military-compiles-biometric-database-on

-iraqi-fighters-saddam-loyalists/d/d-id/1018572?. It is probable from the context that Barnett means collection of biometric data from all possible sources, not only from individuals who have been shown to have ties to insurgent groups, and is not using in its strict sense the technical term "passive collection," which refers to biometric technologies that do not require active participation or even awareness of the individual being scanned.

17. Biometrics Identity Management Agency, *Annual Report, FY10* (Washington, DC: Biometrics Identity Management Agency, 2011), 1, available https://web.archive.org/web/20110719090045/http://www.biometrics.dod.mil/Files/Documents/AnnualReports/fy10.pdf.

18. Noah Shachtman, "Detainees Fill Pentagon DNA Databank," *Wired*, October 14, 2008, available at http://www.wired.com/2008/10/detainees-fill.

19. See Stephen Graham, *Cities under Siege: The New Military Urbanism* (London: Verso, 2010).

20. "Bush Campaign to Base Ad on Kerry Terror Quote," *CNN*, October 11, 2004, available at http://www.cnn.com/2004/ALLPOLITICS/10/10/bush.kerry.terror. Somehow, U.S. abdication of its role as a superpower is never considered as included in "whatever it takes."

21. Keith L. Shimko, *The Iraq Wars and America's Military Revolution* (New York: Cambridge University Press, 2010).

22. There is a good deal of debate within the military community over the role of doctrine and its actual importance to the conduct of war. See, for example, Adam Elkus, "Do Ideas Matter? A Clausewitzian Case Study," *Small Wars Journal*, January 27, 2010, available at http://smallwarsjournal.com/jrnl/art/do-ideas-matter; and Paul Johnston, "Doctrine Is Not Enough: The Effect of Doctrine on the Behavior of Armies," *Parameters* 30 (2000): 30–39. Jim Lewis of the Center for Strategic and International Studies said that these types of rhetorical paradigms "shift every four to five years" and that programs, tactics, and funding simply have to shoehorn themselves into the changed rhetoric regardless of their actual applicability. He cited Woodward's above-mentioned "identity dominance" as one such rhetorical embellishment. (Jim Lewis, interview by the author, February 4, 2011, Washington, DC.) While I absolutely do not want to attribute a simple, deterministic relationship between military doctrine and the development and deployment of biometric technologies, shifts in rhetoric signal the relative political strength of particular factions and philosophies within the Pentagon. This in turn affects funding priorities, troop deployments, and battle tactics.

23. William A. Owens and Edward Offley, *Lifting the Fog of War* (Baltimore: Johns Hopkins University Press, 2001); Shimko, *The Iraq Wars and America's Military Revolution.*

24. U.S. Army and Marine Corps, *Counterinsurgency Field Manual* (Chicago: University of Chicago Press, 2007). See also, David Galula, *Counterinsurgency Warfare: Theory and Practice* (Westport, CT: Praeger Security International, 2006); and David Kilcullen, *Counterinsurgency* (Oxford: Oxford University Press, 2010).

25. Shimko, *The Iraq Wars and America's Military Revolution*, 185.

26. Ben Riley, principal deputy to the deputy assistant secretary of defense, explicitly made the connection between counterinsurgency and biometrics during remarks at an event on the development of Rapid DNA technology by the

DoD, DHS, and the Department of Justice. Ben Riley, opening remarks at the Center for Stategic and International Studies "U.S. Department of Defense Biometric and Forensic Technology Forum," Washington, DC, March 26, 2012.

27. Quoted in Kilcullen, *Counterinsurgency*, 145.

28. Ibid., 45.

29. Galula, *Counterinsurgency Warfare*, 82.

30. John C. Torpey, *The Invention of the Passport: Surveillance, Citizenship, and the State* (Cambridge: Cambridge University Press, 2000), 244.

31. Ibid., 10.

32. See, for example, James C. Scott, John Tehranian, and Jeremy Mathias, "The Production of Legal Identities Proper to States: The Case of the Permanent Family Surname," *Comparative Studies in Society and History* 44, no. 1 (2002): 4–44; and Götz Aly, Karl Heinz Roth, Edwin Black, and Assenka Oksiloff, *The Nazi Census: Identification and Control in the Third Reich* (Philadelphia: Temple University Press, 2004).

33. Associated Press, "Marines Seize Taliban HQ, IDs, Photos," *NBC News*, February 19, 2010, available at http://www.msnbc.msn.com/id/35487636/ns/world_news-south_and_central_asia/.

34. Dale Andrade and James Willbanks, "Cords/Phoenix: Counterinsurgency Lessons from Vietnam for the Future," *Military Review*, March–April 2006, pp. 9–23.

35. See, for example, Roberto J. Gonzàlez, "Towards Mercenary Anthropology? The New US Army Counterinsurgency Manual *FM 3-24* and the Military-Anthropology Complex," *Anthropology Today* 23, no. 3 (2007): 14–19; Sharon Weinberger, "Anthropology Ass'n Blasts Army's 'Human Terrain,'" *Wired*, November 7, 2007, available at http://www.wired.com/2007/11/anthropology-as; and Jason Motlagh, "Should Anthropologists Help Contain the Taliban?" *Time*, July 1, 2010, available at http://content.time.com/time/world/article/0,8599,2000169,00.html.

36. "In those days Caesar Augustus issued a decree that a census should be taken of the entire Roman world" (Luke 2:1 [New Jerusalem Bible]). At the time, the "entire Roman world" was primarily occupied territory. This is not to say that identification in a much larger sense was absent from historical conflict, as mutilation, branding, tattooing, clothing, and naming practices were all used to locate and make legible dangerous populations. Identification cards themselves are a product of modernity and therefore not used in premodern conflict. See Torpey, *The Invention of the Passport*.

37. David Lyon, *Identifying Citizens: ID Cards as Surveillance* (Malden, MA: Polity, 2009); Jane Caplan and John C. Torpey, *Documenting Individual Identity: The Development of State Practices in the Modern World* (Princeton, NJ: Princeton University Press, 2001).

38. Elia Zureik, David Lyon, and Yasmeen Abu-Laban, eds., *Surveillance and Control in Israel/Palestine: Population, Territory and Power* (New York: Routledge, 2011).

39. Helga Tawil-Souri, "Orange, Green, and Blue: Colour-Coded Paperwork for Palestinian Population Control," in *Surveillance and Control in Israel/Palestine: Population, Territory and Power*, ed. Elia Zureik, David Lyon, and Yasmeen Abu-Laban (New York: Routledge, 2011), 229.

40. Bing West, a former Marine and coauthor of the *New York Times* op-ed cited earlier, stated that "the lack of a biometric ID card [is] the greatest technical failure of the Iraq war." Lionel Beehner, "Backgrounder: A National ID Program for Iraq?" *New York Times*, May 29, 2007, available at http://www.nytimes.com/cfr/world/slot2_20070529.html.

41. Michael Dillon and Julian Reid, *The Liberal Way of War: Killing to Make Life Live* (London: Routledge, 2009).

42. Thomas Pratt, interview by the author, Crystal City, Virginia, April 26, 2011.

43. Francis Galton, *Finger Prints* (Buffalo, NY: W. S. Hein, 2003), 94.

44. Alondra Nelson, "Bio Science: Genetic Ancestry Testing and the Pursuit of African Ancestry," *Social Studies of Science* 38 (2008): 759–783.

45. Early fingerprinting experts attempted to shore up their legitimacy in the court system, yet in doing so they transformed their profession into one of technicians rather than scientists. Simon A. Cole, *Suspect Identities: A History of Fingerprinting and Criminal Identification* (Cambridge, MA: Harvard University Press, 2001).

46. Helen Fay Nissenbaum, *Privacy in Context: Technology, Policy, and the Integrity of Social Life* (Stanford, CA: Stanford Law Books, 2010), 4.

47. Ibid. (emphasis in original).

48. Defense Science Board, "Report of the Defense Science Board Task Force on Defense Biometrics," March 2001, available at http://www.dtic.mil/docs/citations/ADA465930.

49. Velliquette, "The Role of Biometrics in the Counterinsurgency."

50. Woodward, Orlans, and Higgins, *Biometrics*; U.S. Central Command, "Concept of Operations for Biometrics in the US Central Command AOR," March 2007, available at https://info.publicintelligence.net/USCENTCOM BiometricsCONOPS.pdf. While the personal data of U.S. soldiers is granted much greater legal protections, their collection is no less compulsory and in fact may be even more intrusive. This is in line with Torpey's description of the correlation between the state's grasp and its penetration. For better or worse (arguably primarily for better), U.S. soldiers are tightly grasped by the U.S. government and thus are deeply penetrated by mediating documents and technologies; see Torpey, *The Invention of the Passport*. The ability of U.S. forces to collect biometrics "at gunpoint" has ceased, or at least become officially improper, following the reinstatement of sovereignty to Iraq. There is now a bilateral agreement between the United States and the Iraqi government governing the collection and sharing of such data. As Afghanistan was considered a sovereign country since soon after the expulsion of the Taliban, a bilateral agreement has existed for almost the entirety of the occupation, thus greatly complicating the collection of such data and at times causing a strain on United States–Afghan relations.

51. Air Land Sea Application Center, "Biometrics: Multi-service Tactics, Techniques, and Procedures for Tactical Employment of Biometrics in Support of Operations," April 2014, pp. 22–23, available at http://armypubs.army.mil/doctrine/DR_pubs/dr_a/pdf/atp2_22x85.pdf.

52. Defense Science Board, "Report of the Defense Science Board Task Force on Defense Biometrics."

53. This was the case in Rwanda, where familial relations often trumped official ethnic classification. See Timothy Longman, "Identity Cards, Ethnic Self-Perception, and Genocide in Rwanda," in *Documenting Individual Identity: The Development of State Practices in the Modern World*, ed. Jane Caplan and John C. Torpey (Princeton, NJ: Princeton University Press, 2001), 345–357.

54. For example, one BIMA annual report (2011) has a section titled "Watching for Bad Guys," although the official terminology has recently started to shift to separating "friend from foe." Biometrics Identity Management Agency, *Annual Report, FY11*, 11, 19.

55. See, for example, Myra Gray, "Director's Message," in *Biometrics Task Force Annual Report FY09* (Washington, DC: Department of Defense, 2010), available at http://fas.org/man/eprint/biometric09.pdf.

56. Biometrics Identity Management Agency, *Annual Report, FY11*, 19. The promotional video is available at https://web.archive.org/web/20120622004611/http://www.biometrics.dod.mil/default.aspx.

57. U.S. Central Command, "Concept of Operations for Biometrics," 5.

58. Ibid., 6. The graphic on page 6 of "Concept of Operations for Biometrics" was shown to me at BIMA as a regular part of their outreach communications.

59. Anup Shah, "World Military Spending," *Global Issues*, June 30, 2013, available at http://www.globalissues.org/article/75/world-military-spending.

60. Galula, *Counterinsurgency Warfare*, 82.

61. James C. Scott, *Seeing Like a State: How Certain Schemes to Improve the Human Condition Have Failed* (New Haven, CT: Yale University Press, 1999), 3.

62. U.S. Central Command, "Concept of Operations for Biometrics," 6.

63. Myra Gray, speech to troops about biometrics, Defense Forensics and Biometrics Agency podcast, June 1, 2009, available at http://www.biometrics.dod.mil/PublicAffairs/podcast.aspx.

64. Gerard Noiriel, "The Identification of the Citizen: The Birth of Republican Civil Status in France," in *Documenting Individual Identity: The Development of State Practices in the Modern World*, ed. Jane Caplan and John C. Torpey (Princeton, NJ: Princeton University Press, 2001), 30–31.

65. Kilcullen suggests that commanders be well versed in the local politics, geography, and tribal and community affairs of their given region such that they are able to determine loyalties, pressure points, and actualizable goals—all within twelve–fifteen months. This seems like a tall order even for a multiyear team of anthropologists. Kilcullen, *Counterinsurgency*.

66. I was told of one bored soldier who had tired of the time and energy it took to register Iraqis and used the shortcut of registering each *with his own fingerprints and iris scan*. By the time his neglect was discovered, he had registered over a thousand individuals with his biometrics. Each entry had to be carefully found and removed from the database.

67. "Lexicon: Clean-Skin Terrorist," *Time*, April 12, 2007, available at http://www.time.com/time/magazine/article/0,9171,1609777,00.html.

68. U.S. Government Accountability Office, "Defense Management: DOD Can Establish More Guidance for Biometrics Collection and Explore Broader Data Sharing," October 2008, available at http://www.gao.gov/assets/290/282770.pdf.

69. Joseph N. Pato and Lynette I. Millett, eds., *Biometric Recognition: Challenges and Opportunities* (Washington, DC: National Academies Press, 2010); Ellen Messmer, "National Research Council Report on Biometrics Raises Hard Questions, Ire," *Network World*, September 28, 2010, available at http://www.networkworld.com/article/2219867/malware-cybercrime/national-research-council-report-on-biometrics-raises-hard-questions--ire.html.

70. Pato and Millett, *Biometric Recognition*, viii.

71. Ibid.

72. R. N. Rodrigues, N. Kamat, and V. Govidaraju, "Evaluation of Biometrics Spoofing in a Multimodal System," in *2010 IEEE Fourth International Conference on Biometrics: Theory, Applications and Systems* (Arlington, VA: IEEE, 2010).

73. Biometrics Identity Management Agency, "Biometrics: Impacting the Warfighter and Beyond," July 2010, available at https://web.archive.org/web/20110427030616/http://www.biometrics.dod.mil/podcast.aspx?.

74. William Bogard, *The Simulation of Surveillance: Hypercontrol in Telematic Societies* (Cambridge: Cambridge University Press, 1996); Gates, *Our Biometric Future*, 6.

75. Michel Foucault, *Discipline and Punish: The Birth of the Prison*, 2nd ed. (New York: Vintage Books, 1995), 195.

76. Graham, *Cities under Siege*, 146–147.

77. Foucault, *Discipline and Punish*, 200.

78. Ibid., 202.

79. Oded Na'aman, "The Checkpoint," *Boston Review*, November 13, 2012, available at http://bostonreview.net/world/checkpoint-oded-naaman.

6 /

Precision Warfare and the Case for Symmetry

Targeted Killings and Hostage Taking

Ariel Colonomos

For more than a decade, wars fought by Western states—mostly by the United States, the United Kingdom, and Israel—have been labeled "asymmetrical."[1] The shift from conventional toward asymmetrical warfare has been characterized as a key feature in the contemporary transformation of war. The model of asymmetrical warfare opposes two sets of actors that are divided along economic as well as political and organizational lines. It is a duel between the rich and the poor, modern states equipped with the latest military technology fighting either weak states or nonstate actors that lack sophisticated weaponry. It entails a confrontation between, on the one hand, hierarchical and bureaucratized states and, on the other, fragile states or violent nonstate actors that challenge the principle of state sovereignty. The model also has a third dimension. In normative terms, the idea of asymmetrical warfare entails a difference between democratic states where the rule of law prevails and that attempt to abide by international humanitarian law (IHL) and other states or nonstate actors that refuse to abide by those norms.[2]

This chapter argues against the notion that these conflicts are asymmetrical. It points out the simplifications that this model entails, which stem from a conceptual lack of robustness resulting, in certain cases, from ideological biases. I highlight conditions of symmetry in those situations that are portrayed as asymmetrical. Some literature in law and ethics already criticizes the asymmetrical model, including its normative thesis, according to which defenders of IHL, in this case democracies, are in inevitable opposition to its transgressors. Indeed, terrorist organizations are not the only entities that kill civilians, and democratic states also have a long-standing tradition of killing noncombatants.[3]

This chapter defends the case for a symmetry model. It does so in a quite different manner than the mainstream objection against it, which relies on assumptions about the lack of precision for both parties involved in so-called asymmetrical warfare. I argue that precision is, on the contrary, a common tactic that both parties, strong democratic states and nonstate actors alike, use in their contemporary military operations.[4]

The main argument of the chapter concerns the structure of warfare in contemporary international politics. Although it refers to the debates about targeted killings in the just-war framework and the killings' legality, this chapter is not an analysis of the ethics of war or the contemporary use of force, and it does not morally equate targeted killings to hostage taking. Rather, the goal is to question the notion of asymmetrical warfare and raise doubts about the implicit consensus and the illusions on which that consensus is premised. Words matter; misuse of words is morally misleading, supports bad decisions, and limits the democratic lawmaking process.

Warfare has greatly developed over the last fifteen years. Two parallel phenomena, targeted killings and hostage taking, are highly characteristic of this evolution. Targeted killings have become routine for some countries, such as Israel and the United States. More recently, hostage taking has become one of the main modes of confrontation that groups such as Al Qaeda, Hamas, or Daish (also known as the Islamic State or ISIS, a label I reject) use with their enemies—namely, the United States, Israel, the United Kingdom, and their allies.

This chapter argues that targeted killings and hostage taking are two important symmetrical features of contemporary warfare. I first show that these two sets of practices are constitutive of the age of precision and one of the most important generic characteristics of contemporary warfare. I then argue that both targeted killings and hostage taking are violations of international law and are highly subversive of both the political framework of the Westphalian order and the model of the just-war tradition. In the third part of the chapter, I contend that these two practices reflect similar sets of beliefs and values on the part of those who are engaged in such activities and that those beliefs and values are contrary to the ideal and traditional ethos of the art of war. Finally, I ask whether targeted killings and hostage taking reinforce each other and contribute to a new type of escalation in the dynamic of violence.

The Age of Precision

The history of aerial bombing reflects major strategic, political, and moral choices. It shows the capabilities of nation-states engaged in combat and their armies, the intention of their leaders, and the outcomes of their decisions. Aerial bombing, as it has evolved in the twentieth century, gained in precision, and indeed, the change has been quite significant.[5] In the first part of the twentieth century, Western armies did not hesitate to follow the doctrine initiated between the two world wars by Italian general and air power theorist Giulio Douhet, according to whom terror bombing was the key to a rapid victory.[6] During World War II airpower was used by both sides to bring down the morale of entire nations whose civilians were dying in the rubble of their homes. Indeed, in air warfare, democracies showed no more mercy than their totalitarian counterparts.[7]

The policy of intentional killing of civilians changed after World War II.[8] The Vietnam War was a turning point, as the United States started to alter its policy of massive bombings. The role of military lawyers in the army was greater in the Vietnam conflict than it had been previously.[9] Later in the twentieth century, the laws of war occupied a larger role in planning airstrikes.

Technological developments have made precision possible. Laser-guided weaponry and GPS have served as crucial enablers of targeted killings.[10] New beliefs and assumptions about those who fight and respond to fire have guided decisions about the need for precision as well as the investments in technology it has necessitated. According to the Douhet framework, the morale of nations is hurt if civilians are killed in massive numbers. The more contemporary assumption about the behavior of those whose relatives or fellow nationals are killed during a war, however, is quite the opposite. It is today often assumed that civilians dying in large numbers would leave combatants' willingness to fight undiminished or even enhanced.[11]

Given the moral and legal pressures with which Western armies are confronted, precision is a response to challenges by lawyers, nongovernmental organizations, or other states that might want to engage in legal action in pursuit of war crimes or human rights violations. It is also, to a certain extent, a preemptive measure of precaution, as these critiques are well anticipated. "Naming and shaming" has become routine for nongovernmental organizations, and "lawfare"—the legal battle that parallels the military battle on the ground when states and soldiers are being accused of IHL violations—has considerably expanded over the last twenty years.[12]

The move toward precision accelerated in the 1990s. During the first Gulf War, the United Nations coalition used some precision-guided weaponry to fight Saddam Hussein. During the Kosovo campaign, precision became an even bigger concern. When fighting Slobodan Milošević, the North Atlantic Treaty Organization (NATO) was confronted with a dilemma arising from the need to protect its pilots and its simultaneous desire to minimize the number of civilian casualties. NATO planes flew at a high altitude so as to protect the pilots, yet the number of civilians who died from the bombings was relatively small compared to previous air campaigns.

Emphasis on the precision of strikes that are aimed at specific individual targets takes place in a context in which human rights activists are challenging Western states and in which the killing of civilians spurs criticism. From this perspective, Israel has been confronted with this new emphasis on the protection of civilians in the context of international conflict. Shin Bet (the domestic security forces) introduced this new idea, and Israel framed its new policy of targeted killings (*sikul memukad* in Hebrew, literally "preventive elimination") in 2000 when the Second Intifada began. Although it adopted a similar tactic as a reaction to the 9/11 attacks, the United States initially criticized the Israeli policy of targeted killings.

Targeted killings can be seen as the continuation of the evolution of aerial bombings toward an expectation of ever-greater precision.

Precision serves not only as constraint but is also an objective on the part of Israel and the United States. Both believe that killing certain individuals—in the case of the United States, high-ranking officials of organizations such as Al Qaeda and, in the case of Israel, usually midlevel Palestinian combatants who are seen as the future leaders of Hamas and other militant Islamist factions—will deeply alter the functioning of these groups, prevent future attacks, and ultimately cause these organizations to crumble.

The wars in Afghanistan and Iraq marked a new step in the evolution of warfare. Western troops, mostly from the United States and the United Kingdom, were on the ground and engaged in wars that lasted much longer than those who planned them initially thought they would. This gave the opportunity to the groups the United States was fighting, mostly Al Qaeda and other Islamist movements, to develop hostage taking as a policy to confront and defy Western forces. Hostage taking has a long history in the Middle East and in Africa and, during the war in Lebanon in the 1980s, several Westerners had been held hostage by Hezbollah. For the first time, however, the abduction of both civilians and military personnel became central in the lives of transnational armed movements that were challenging the brokers of international order.

Al Qaeda is most remembered for the spectacular atrocity of the attacks against the World Trade Center and the Pentagon. However, through its affiliates in the Maghreb and in Africa, it has developed a new strategy in which kidnapping is one essential element. Abduction necessitates extreme precision. In that respect, hostage taking radically differs from indiscriminate terrorist attacks such as those carried out in New York and Washington in 2001. Quite disturbingly, it echoes targeted killings. Both require information about the targets that are either to be eliminated or kidnapped (although sometimes abductions are not premeditated and depend on the encounter of radical Islamist groups with potential hostages). Both are well acknowledged by their perpetrators. Indeed, targeted killings are part of the official security doctrines of Israel and the United States. Israel discloses the names of those it kills after each operation. American policy is less transparent, but the logic of drone operations and their legal rationale are openly debated in the media and in academia.[13] Hostage taking is an essential element in the communication that militant Islamist groups want to establish with their enemies. Abductions are usually fully recognized by the organizations that are responsible for them.

Both practices require extensive use of technology. Targeted killing operations often require the use of drones, precision-guided missiles, or in some instances, microexplosive devices inserted into cell phones used by the targets; moreover, these operations are also sometimes filmed and the Israeli Defense Force (IDF) broadcasts the footage as examples of successful operations that hit their targets without killing civilians. Hostage taking necessitates the use of the media and the Internet to communicate with the family members of those who have been abducted and, more generally, with their leaders, fellow citizens, and other relevant publics. Daish's recent Internet videos of beheadings are the latest macabre development of this evolution. Eventually, nonstate actors may rely on easily accessible drone technology to locate potential hostages.

Unlawful Practices

Precision is a normative requirement in warfare. Yet in itself, it does not make the use of force legitimate and legal, as other criteria must be met to meet these standards. In practice, however, precision has been used as a rationale for condoning, if not justifying, the use of force. Concepts such as the humanization of warfare or clean wars have proliferated and are a reflection of the evolution of warfare and its perception.[14] Western powers use precision as a mode of justifying the use of force in wars that are fought over dubious causes, especially in the case of preventive wars. States are looking for a moral luck effect—if a war is fought properly and does not involve too heavy a burden for the target's civilian population, it will be perceived less negatively than if it had caused heavy suffering and the laws of distinction and proportionality had been violated. There lies a paradox here, as this preference for precision and distinction has one major side effect: it has had made it easier for states such as the United States and Israel to wage wars.[15]

Targeted killings have fueled interesting and controversial debates on the laws of war.[16] One of the first indicators of their controversial nature can be found in the harsh disputes over the naming of this practice. Indeed, "targeted killings" is the most consensual phrasing employed to designate these operations. Others include "targeted assassinations" or, in some countries, "targeted attacks." On the other side of the spectrum, one of the more euphemistic—one might even say Orwellian—terms is Israel's "preventive eliminations."

In academia, partisans of this practice have argued that the measure is ethical or lawful or both.[17] In the United States, Obama administration lawyers in the military and at the State Department have also tried to convince the wider public that targeted killings are a legitimate response to the evils of global terrorism.[18]

Targeted killings are disturbing because, by definition, they affect individuals in contexts that fall outside the traditional realm covered by the just-war tradition and the dictates of international law—the realm of interstate war. The purpose of the just-war tradition has been to establish rules regulating the behavior of states in order to avoid wanton destruction, minimize suffering, and establish the pillars of an international society of states in which civility and caution prevail. Although in some respects it is different and more restrictive in the limitation of the use of force it defines, international law pursues similar objectives. Both in the just-war tradition and in international law, targeted killings have never been properly identified and discussed as such, which is one of the reasons they have raised so much interest over the last fifteen years and have fueled so many controversies.

There have been a number of serious critiques of the practice of targeted killings. I argue here that, overall, the practice cannot be justified on legal grounds and, as a general doctrine, is not morally acceptable. The most significant shortcoming of targeted killings concerns the violation of sovereignty that the practice entails. In a great many instances when the United States uses drones to attack its opponents, the operation takes place in countries with which the United States is not at war, such as Yemen or Pakistan. The violation of sovereignty is well characterized. Some would argue that, in certain instances, foreign leaders have authorized the U.S. military to conduct its mission on their soil. This authorization, however, does not make these killings a lesser violation of sovereignty. A state is an organization that has the monopoly over the use of force on its territory. This Weberian political definition of the state fuels a vision of sovereignty according to which, indeed, no state or private actor is allowed to make use of force on another state's territory. In the case of targeted killings, the government of Yemen or Pakistan can secretly authorize the United States to strike in its territory. However, from the perspective of sovereignty, one must differentiate a government from the state. A governmental decision can hurt the state; in this case, a governmental decision hurts the principle of state sovereignty. Moreover, targeted killings affect the very principle of the state. The United States frequently laments that weak states are a threat to security worldwide. We might see here a contradiction, as this measure seems to weaken even more those fragile state entities.

The Israeli case is more specific, given that Israel targets militants who do not belong to a state and who fight on fragmented soil regulated by different authorities.[19] Lawyers have argued extensively about which law should apply to the discussion concerning the legality of targeted killings. Aaron Barak, the well-respected Israeli lawyer and former president of the Israeli Supreme Court, argues that when fighting in conflicts—that is, bearing arms and making use of them—civilians lose their immunity and therefore become legitimate targets of *preemptive* strikes when engaged in violent action, at least under certain conditions.[20] Some, such as David Kretzmer, argue that targeted killings are a severe violation of human rights law, even though they acknowledge that, if IHL were to be applied, targeted killings would be lawful if those fired on are combatants, the principle of distinction applies, and the number of civilians who die as collateral damage is limited.[21] Because targeted killings happen in territories that are not theaters of interstate warfare, which legal regime applies to the regulation of the use of force is controversial. IHL might not apply, but in any event, human rights law should apply, as it is valid in war and peace.

Especially in the Israeli case, targeted killings are justified on the basis of prevention as part of the *sikul memukad* doctrine.[22] Since 9/11, prevention has become one of the main justifications of military intervention and, more generally, the use of force.[23] Leaving aside the inherent legal problems of justifying prevention, in the context of American operations more specifically, the information given by the United States about the nature of the targets (both ex ante and ex post) is scarce or nonexistent. Thus, the case for prevention is all the more difficult to establish, and the U.S. government remains unaccountable.

As a policy, targeted killings are dangerous to the extent that they may violate and endanger an established and important norm: the rule that bans the assassination of foreign political leaders.[24] Indeed, although targeted killings are said to target combatants in times of conflict, they sometimes take place in countries with which the United States is not at war. Such action violates the rule that bans the assassination of political leaders, which was implemented precisely to prevent the political troubles and instability that result from the assassination of political leaders by foreign powers in the absence of war.[25] Authorizing targeted killings in the absence of international conflict against individuals who are combatants and, for some at least, political leaders within their own organizations can therefore jeopardize the existing norm.

Morally, and in very specific cases, targeted killings might be justifiable as an exceptional measure, especially when they preempt an incoming attack that would harm American civilians. However, as a norm and a doctrine, this policy is damaging to the international legal structure and to the international normative order. Individual decisions may be justifiable from the perspective of act utilitarianism, but the policy fails the test of rule utilitarianism.[26] A single targeted killing may be justified if it enhances the security of the United States and protects it from an incoming attack. However, the repeated use of this tactic—a *doctrine* of targeted killings—is damaging from the perspective of rule utilitarianism, as it endangers a norm that has proved useful: the ban on the killing of foreign political leaders. We see other negative consequences that result from the targeted killings doctrine, as in Israel's systematic use of Palestinian informants and the sense of distrust that this creates in Palestinian society.[27]

The case against contemporary hostage taking is more straightforward both legally and ethically, although in ancient times the Greeks and more particularly the Romans took hostages, and this practice was considered to be a means to reach a peace settlement. Hostages were often the sons of foreign political leaders and were treated with great civility. They received a Roman education, and some became Roman citizens.[28] They served as go-betweens and mediators between the empire and its enemies, and they spread the influence of Roman culture.

However, hostage taking in warfare has changed over time.[29] In the twenty-first century, hostages fail to be treated humanely, and they no longer serve as a means to find agreements in conflicts between those who kidnap soldiers and civilians and those countries whence the abducted originate.

Hostage taking is now considered a clear case of violation of both IHL and human rights law. According to IHL, states must treat prisoners according to the rules established by the Geneva Conventions. They must receive humane treatment, and humanitarian organizations must have access to them. Prisoners are to be set free at the end of the conflict. However, in the case of a conflict between states and nonstate actors, these conditions are hardly likely to be met, especially on the part of nonstate actors that use those whom they take prisoners as hostages—essentially, they are treated as bargaining chips in the negotiations with powerful state adversaries. Some would even challenge the right of nonstate actors to take prisoners.[30]

As for human rights law, hostage taking is a strong violation of this legal regime. Hostage taking violates the rights of civilians, exposing them to undue suffering. For some who argue that human rights law should also apply to soldiers, hostage taking becomes a violation of the rights of persons to be treated in a dignified way. Moreover, in 1979 the Convention against the Taking of Hostages made the practice a clear violation of international law.[31] It is also considered a crime in some domestic jurisdictions; it is, for example, a crime according to U.S. federal law.[32]

Morally speaking, hostage taking causes undue suffering. It is also a radical form of blackmail and uses human beings as means to pursue an end, which makes it—from a deontological perspective—impossible to justify. The consequences of hostage taking are also nefarious, as this practice rather often drives states into armed conflicts. For example, United States–led coalition forces openly moved to use force against Daish after it took several hostages and beheaded them. The consequences are all the more negative, because when states retaliate against those who have taken their nationals hostage and try to free them in urban areas, where they tend to be hidden, those states may be led to violate the rules of proportionality in war and cause suffering among the populations where hostages are thought to be hidden.

Two Converging Types of Ethos

It may seem odd to compare the ethos of those whose stated ambitions are to make the world safe for democracy and to resist terrorism with that of those whose declared goals are to establish sharia law, kill infidels, and establish a caliphate. However, disturbing similarities between the practices of the two groups should lead us to rethink the status of precision warfare in democracies.

Both targeted killings and hostage taking have a strong punitive dimension and are likely to be motivated by revenge. Although the former, especially in the Israeli case, are officially justified by the need to take preventive action in the face of future attacks by terrorists (and therefore, the criteria for selecting the targets is not primarily the past behavior of the person to be killed), these military operations are often perceived in Israeli society as just retribution when they strike combatants who have already been involved in terrorist operations that killed civilians. In the Israeli case, the identity of the target as well as the role he or she played in one of the Palestinian organizations that Israel wants to weaken are disclosed to the public. Especially compared to the behavior of other democracies that operate in secret, Israel has a transparent policy. However, the public does not have substantial information about the nature of the threat posed by the person whose life is taken. Evidence about whether the *sikul memukad* policy led to positive outcomes in fighting terrorism is lacking. Targeted killings were not decisive in ending the suicide attacks that characterized the Second Intifada. The ending of suicide bombing attacks has instead been the consequence of the wall erected to separate Israel from Palestinian territories.[33] When Daish takes a hostage, punishing the hostage is one of the main reasons for the abduction. Outfitting hostages in orange jumpsuits, just like those worn by prisoners at Guantánamo, signals that executions, such as beheading the hostage, are punishment for the crimes of those countries from which the abducted originate and their allies. From this perspective, they are a spectacular form of collective punishment.

Virtue ethics is a philosophical tradition that comes from the Greeks and the Romans and praises the noble behavior of those whose deeds are in accordance with virtues such as temperance or courage.[34] Virtue ethics serves as an alternative framework for the ethics of war in which the just-war tradition prevails and in which Kantianism and utilitarianism are the dominant philosophical referential concepts.[35] Ethics is being taught in military academies in Western countries, and virtues such as temperance, courage, discipline, and sacrifice occupy a central role in those teachings.[36] These foundational values are both a model for the building of an ethos of warfare and a reflection of this shared military ethos. Some would highlight the deep ethical values and the stoic virtues that characterize military personnel.[37] From this perspective, both targeted killings and hostage taking appear as strikingly opposite to these values and stand in stark contradiction to the very idea of the ethics of virtue. Targeted killings have been criticized for perverting the art of war, because such practice resembles more a manhunt than anything else. Hostage taking is, of course, even more a perversion of virtue ethics in warfare, as both military personnel and civilians are abducted by hostage takers.

Although we do not have extensive data on the psychological reactions of the military who are involved in targeted killings, some of the military personnel who operate U.S. drones reportedly suffer psychological problems related to their task.[38] Insofar as this is the case, it would stand as evidence of the contradictions or the moral dissonance between these practices and the values and ideals that are proclaimed by the military and to which many soldiers subscribe.

Israeli soldiers' reactions to targeted killings are extremely diverse.[39] Some soldiers feel shame, remorse, or pain. Others consider targeted killings part of their duties while serving in the army in the context of international conflict. However, we also have examples of other soldiers who take pride in performing targeted killings and who express their satisfaction at mastering precision shooting. Pride is also a reaction that, in its most extreme and horrifying version, we find in the footage disseminated by Daish when one of its members beheads one of their victims. Despite the magnitude of the offense and the disgust provoked in the latter case, targeted killings and hostage killings share a problem: the pleasure felt when harming an unarmed person.

Both targeted killings and hostage taking imply the use of deception, notably when searching for the specific person to kill or abduct. In the just-war tradition, deception is, to a certain extent, permissible. However, in these two cases, ruses deceive those who are unarmed and, as they are not professional soldiers, less protected and more exposed to the dangers of deception. Even in targeted killings, combatants are, in a great many instances, targeted at home and, more generally, going about their everyday lives. In hostage taking by Daish or Boko Haram, those who are kidnapped have not usually taken part in combat and are often journalists.

The Israeli policy of targeted killings is praised for its precision. Indeed, the number of civilians who are killed unintentionally in those operations is much lower than in regular warfare.[40] The reason these operations avoid massive casualties lies in the quality of the information that the IDF receives. The IDF gets its information from a large network of informants who collaborate with it. This creates a wide feeling of distrust within Palestinian society, where those who collaborate with Israel are severely punished (they are usually put to death).[41] Peace will be that much more difficult to build for Israel with a fragmented Palestinian society as negotiating partner. The consequences of prolonging this doctrine and this practice while relying on a network of coerced informants are likely disastrous. Similarly, hostage taking relies on information that is gathered by means that are sometimes dubious.

Self-Reinforcing Mechanisms?

Targeted killings and hostage taking share similarities, but do they also reinforce each other? Some facts indicate a nexus between the two practices, but these are mere hypotheses that need to be tested thoroughly.

The expansion of hostage taking coincides with the routinization of targeted killings. Indeed, although hostage taking has long been a very important concern for Israel, the Gilad Shalit case marked an important step in the history of the abduction of Israeli soldiers and the negotiations to free them. Gilad Shalit was captured in 2006 by

Hamas in Gaza and released in exchange for 1,027 Palestinian prisoners five years later. Targeted killings started late in the first decade of the 2000s, and the policy has been developing and expanding. Since then, Hamas and Hezbollah groups have repeatedly called for the abduction of Israeli soldiers.

We might find a relation between the two, as hostage taking can be a way to compensate for the losses created by targeted killings. There is no strong evidence of the efficacy of targeted killings in terms of security. However, without incapacitating the militant Islamist movements they target, these killings temporarily weaken the structure of the organizations whose members are being killed. The group needs to replace the members and train them. In the face of these shortages, hostage taking brings resources; in the case of Palestinian groups, it can lead to the release of a significant number of its members. For Al Qaeda and its affiliates, whose members have been pursued by drones since the early 2000s—the first American operation acknowledged by the Central Intelligence Agency goes back to 2004 in Yemen—abduction brings in economic resources that are indispensable to the survival and development of these groups.[42]

Psychological factors might also come into play in the connection between these two practices. Targeted killings involve attributions of responsibility to agents of hostile actions who are wanted in a global setting. For their part, members of groups such as Al Qaeda might want to attribute responsibility for what they consider to be the crimes perpetrated by the West and organize their own conditions of detention and sentencing. It has been extremely difficult for Al Qaeda to strike U.S. soil and very difficult for Hamas to carry out attacks inside Israel unless it starts an open war and faces retaliation, as it did in mid-2014. Therefore, Al Qaeda and Daish use civilians as proxies in this process of attributing responsibility that is followed by expressions of pride for bringing about public executions that are broadcast all around the world. This might be a very specific case of revenge, in which the reaction to perceived injustices is emulation of the enemy, a mimicking of justice.

More generally, this dualism—targeted killings by strong powers and hostage taking by the pariahs of the international order—reinforces the opposition and the perceived divide between civilization and barbarity. Targeted killings are an unusual use of force that is justified on the basis of the need for exceptional measures to fight those who are stigmatized as unlawful combatants.

Conclusion: Rules of War beyond Asymmetry

The assumptions embedded within the notion of asymmetrical warfare are simplistic and sometimes erroneous. Indeed, the opposition between state and nonstate actors is assumed to be based on an inequality of status, knowledge, and resources. Moreover, asymmetrical warfare is described as a structural system in which democracies respectful of the rule of law oppose unlawful nonstate actors. However, this moral opposition does not hold up empirically. I show that there is, to some extent, a case for symmetry in today's confrontation between democracies such as the United States and Israel and the violent organizations that challenge them, as both states and nonstate actors favor precision when they fight.

The goal of this chapter is to challenge the certainties of established notions that obscure our understanding of war. Without falling into moral relativism and the nihilistic equation of democratic practices to those of Daish, Al Qaeda, or Hamas, this chapter calls for a reform of the targeted killings policy—that is, its doctrine and its justification on the basis of an unwarranted system of legal of justification—and for stopping its systematic use and its glorification as the ultimate remedy for terrorism. It also calls for a wide debate on the rights of hostages and the duties that fall on those whose responsibility it is to free them.[43] Instead of merely denouncing hostage taking as a barbarous act (which it is), we need to define the rights of hostages and the duties of states in the face of this expanding practice.

In the context of this opposition between state action and the challenges of violent nonstate actors, if we fail to distance ourselves from ideologically biased, catchall concepts, we risk the global proliferation of zones of anarchy and local niches of violence.

Notes

1. "Asymmetrical warfare" is a widespread notion that has become diffused in academic circles and among military experts and think tanks, as well as in the media. See, e.g., T. V. Paul, *Asymmetric Conflicts: War Initiation by Weaker Powers* (New York: Cambridge University Press, 1994); and Ivan Arreguin-Toft,

"How the Weak Win War: A Theory of Asymmetrical Conflict," *International Security* 26, no. 1 (2001): 93–128.

2. On the killing of civilians as a severe violation of IHL that should also apply to nonstate actors, see Samuel Eistrecher, "Privileging Asymmetric Warfare: The Intentional Killing of Civilians," *Chicago Journal of International Law* 12 (2009): 589–603.

3. See Michael Gross, "Asymmetric War, Symmetrical Intentions: Killing Civilians in Modern Armed Conflict," *Global Crime* 10, no. 4 (2009): 320–336. Democracies can be criticized for killing civilians even if they do not adopt a policy of targeting noncombatants. Some would then argue that the death of noncombatants, reaching a certain number, testifies to some degree of intentionality on the part of those who have killed them. According to the critics of American policy, in the case of massive numbers of civilian deaths, the usual interpretation of the doctrine of double effect—according to which intending is not foreseeing—would not shield them from accusations of wrongdoing. The number of civilians killed during United States–led drone operations is problematic. First, because many civilians are being killed, these authors put into question that the United States is merely foreseeing their deaths and not intending it. More substantively, because proportionality is one of the criteria of the doctrine of double effect and because civilians are sometimes killed in great numbers (e.g., a dozen civilians in elimination of one target), one may question the morality of this action that is justified on the basis of the doctrine of double effect. This is a very contentious issue in legal and ethical debates. This position does not reflect the common understanding of IHL and the traditional use of the doctrine of double effect. Although this position raises the interesting and complex question of intentionality, I do not agree with it. For an account of the killing of civilians by democracies, see Alexander Downes, *Targeting Civilians in War* (Ithaca, NY: Cornell University Press, 2008).

4. As Michael Gross underlines, both targeted killings and hostage taking are very personal. See *The Ethics of Insurgency: A Critical Guide to Just Guerrilla Warfare* (Cambridge: Cambridge University Press, 2014), 124.

5. Ward Thomas, *The Ethics of Destruction: Norms and Force in International Relations* (Ithaca, NY: Cornell University Press, 2001).

6. Giulio Douhet, *Il dominio dell'aria: Saggio sull'arte della guerra* [The air domain: Essay on the art of war] (1921; repr., Rome: Amministrazione della Guerra, 1932). See also Thomas, *The Ethics of Destruction*, chap. 4.

7. Some argue that they inflicted more pain during their aerial campaigns than Nazi Germany did when it bombed London. Downes, *Targeting Civilians in War.*

8. Thomas, *The Ethics of Destruction*, chap. 5.

9. This parallel between a changed policy of bombings and an increase in the number of military lawyers is also apparent when comparing bombings in Vietnam to bombings in Korea. Ariel Colonomos, *The Gamble of War: Is It Possible to Justify Preventive War?* (New York: Palgrave, 2013), 99–144.

10. Ibid.

11. In its contemporary form, this debate dates to World War II. Chris Jochnick and Roger Normand, "The Legitimation of Violence: A Critical History of the Laws of War," *Harvard International Law Journal* 35, no. 1 (1994): 89.

12. Robert Drinan, *The Mobilization of Shame: A World View of Human Rights* (New Haven, CT: Yale University Press, 2002); Charles Dunlap Jr., "Lawfare: A Decisive Element of 21st-Century Conflicts?" *Joint Force Quarterly* 54 (2009): 34–39; David Luban, "Lawfare and Legal Ethics in Guantanamo," *Stanford Law Review* 61, no. 2 (2008): 1981–2026.

13. Drone warfare also pervades popular culture in the media and in TV shows in Israel and in the United States such as *Hatufim* and *Homeland*. Both fictional shows are about the return of soldiers from their captivity as hostages in the Middle East. Both include scenes of shooting at combatants from Islamist movements. The U.S. policy of drone warfare is at the core of *Homeland*'s season four.

14. Among others, see Christopher Coker, *Humane Warfare* (London: Routledge, 2002).

15. Colonomos, *The Gamble of War*, 99–144. More generally, for the potential negative side effects of the emergence of some norms that, per se, could testify to a certain moral and legal progress, see Richard Price, "Moral Limit and Possibility in World Politics," *International Organization* 62 (Spring 2008): 191–220.

16. See, notably, Claire Finkelstein, Jens Ohlin, and Andrew Altman, eds., *Targeted Killings: Law and Morality in an Asymmetrical World* (Oxford: Oxford University Press, 2012).

17. Steven David, "Israel's Policy of Targeted Killing," *Ethics and International Affairs* 17, no. 1 (2003): 111–126; Asa Kasher and Amos Yadlin, "Assassination and Preventive Killing," *SAIS Review of International Affairs* 25, no. 1 (2005): 41–57.

18. Harold Koh (legal advisor to the U.S. State Department), speech delivered at the 104th annual meeting of the American Society of International Law, Washington, D.C., March 25, 2010.

19. Some lawyers would argue that, under certain conditions, targeted killings could be authorized in areas where the Palestinian Authority is in charge of the regulation of security, whereas it could not in areas where Israel is in charge of security. Targeted killings in the latter areas would signify that the state kills criminals preventively. David Kretzmer, "Targeted Killing of Suspected Terrorists: Extra-judicial Executions or Legitimate Means of Defence?" *European Journal of International Law* 16, no. 2 (2005): 171–212.

20. Public Committee against Torture in Israel et al. v. The Government of Israel et al., HCJ 769/02 (December 14, 2006), available at http://www.haguejusticeportal.net/Docs/NLP/Israel/Targetted_Killings_Supreme_Court_13-12-2006.pdf.

21. Kretzmer, "Targeted Killing of Suspected Terrorists."

22. Kasher and Yadlin, "Assassination and Preventive Killing."

23. Colonomos, *The Gamble of War.*

24. Ward Thomas, "The New Age of Assassination," *SAIS Review of International Affairs* 25, no. 1 (2005): 27–39.

25. Thomas, *The Ethics of Destruction*, chap. 3.

26. Colonomos, *The Gamble of War*, 145–173.

27. Michael Gross, "Fighting by Other Means in the Mideast: A Critical Analysis of Israel's Assassination Policy," *Political Studies* 51, no. 2 (2003): 350–368.

28. Joel Allen, *Hostages and Hostage-Taking in the Roman Empire* (Cambridge: Cambridge University Press, 2011).

29. Adam Kosto, *Hostages in the Middle Ages* (Oxford: Oxford University Press, 2012), chap. 7. One can also think about domestic conflicts and civil wars. Abraham Lincoln wrote in 1863 that "a hostage is a person accepted as a pledge for the fulfillment of an agreement concluded between belligerents." John C. Griffiths, *Hostages: The History, Facts, and Reasoning behind Hostage Taking* (London: André Deutsch, 2003), 13. See also Jon Elster, "Kidnappings in Civil Wars," paper prepared for the Workshop on the Techniques of Violence, Oslo, August 20–21, 2004.

30. Even when prisoners are not necessarily used as hostages, this right should be denied. Tami Meisels, "Kidnapped," in *Soft War*, ed. Michael Gross and Tami Meisels (New York: Cambridge University Press, forthcoming). As opposed to this view, Michael Gross argues that nonstate actors ought to be authorized to take prisoners. These would become hostages, but if they are treated humanely, this practice ought not to be considered as unethical and ultimately illegal. Indeed, it shares with prisoner taking one of its main objectives—weakening the strength of the enemy by neutralizing some of its soldiers. Given how Daish abductees are treated, this would not apply to them. In the case of Hamas, the commodification of the hostage also goes against some basic principles of humanity. An opposition to the instrumentalization of prisoners entails applying this critique to prisoners taken by Israel or the United States when captured for information. Their value lies in the quality of the information they provide. Gross, *The Ethics of Insurgency*, chap. 5.

31. "International Convention against the Taking of Hostages," December 17, 1979, available at http://www.un.org/en/sc/ctc/docs/conventions/Conv5.pdf. According to the convention, any person who takes hostages shall be either prosecuted or extradited: "Any person who seizes or detains and threatens to kill, to injure or to continue to detain another person (hereinafter referred to as the 'hostage') in order to compel a third party, namely, a State, an international intergovernmental organization, a natural or juridical person, or a group of persons, to do or abstain from doing any act as an explicit or implicit condition for the release of the hostage commits the offence of taking of hostages ('hostage-taking') within the meaning of this Convention" (Art. 1).

32. See U.S. Code, Title 18, Chapters 55 and 1203.

33. Daniel Byman, "Do Targeted Killings Work?" *Foreign Affairs*, March–April 2006, pp. 95–111. Edward Kaplan, Alex Mintz, and Shaul Mishal, "Tactical Prevention of Suicide Bombings in Israel," *Interfaces* 36, no. 6 (2006): 553–561.

34. Courage in the face of war is a virtue in Aristotle's thinking. See Aristotle, *Nicomachean Ethics*, bk. 3, chap. 6, 1115a32–35.

35. David Chan, *Beyond Just War: A Virtue Ethics Approach* (New York: Palgrave, 2012).

36. Paul Robinson, "Ethics Training and Development in the Military," *Parameters* 37, no. 1 (2007): 23–36.

37. Nancy Sherman, *Stoic Warriors: The Ancient Philosophy behind the Military Mind* (Oxford: Oxford University Press, 2007).

38. Denise Chow, "Drone Wars: Pilots Reveal Debilitating Stress beyond Virtual Battlefield," *Live Science*, November 5, 2013, available at http://www.livescience.com/40959-military-drone-war-psychology.html.

39. For interviews with Israeli snipers, see the documentary by Nurit Kedar, *One Shot* (2004).

40. From 2001 to 2007, on average, each time the IDF reached its target, one civilian was unintentionally killed (a level of collateral damage of 0.5). Colonomos, *The Gamble of War*, 157.

41. Gross, "Fighting by Other Means in the Mideast."

42. Ransoms paid by Western states or companies are an essential part of the financing of Al Qaeda and its affiliates in Africa. See Rukmini Callimachi, "Paying Ransoms, Europe Bankrolls Qaeda Terror," *New York Times*, July 29, 2014, available at http://www.nytimes.com/2014/07/30/world/africa/ransoming-citizens-europe-becomes-al-qaedas-patron.html.

43. See Ariel Colonomos, "The Hostage Dilemma: The Obligation to Act Globally," in Gross and Meisels, *Soft War*.

7 /

Militarizing Ethnography

The Pentagon's Use and Abuse of Culture

ROBERTO J. GONZÁLEZ

This chapter charts some of the ways Pentagon planners have attempted to incorporate culture into U.S. Department of Defense initiatives. Military officials have paid increasing attention to cultural and ethnographic knowledge over the past decade, and in so doing they have transformed war preparation and war fighting. The shift began slowly, confined mostly to discussions at the U.S. Army's Combined Arms Center at Fort Leavenworth, Kansas. But in recent years, top DoD officials have revised field manuals, issued mandates, and created policies, programs, and training centers, signaling an abrupt change in the way ethnographic knowledge is valued and used by the military. After a long hiatus, culture has returned to the Pentagon.

The renewed interest in cultural knowledge coincided with the rise of a "small band of warrior-intellectuals" in the post-Rumsfeld era, led by U.S. Army general David Petraeus.[1] Petraeus, who has a doctorate in international relations from Princeton University, served in Iraq from 2003 to 2008 in various capacities, including as commander of the 101st Airborne Division, of the Multi-National Force–Iraq, and of the U.S. Central Command. He assembled a team of social scientists with doctoral degrees in history and the social sciences who rose to prominence as the Bush administration desperately sought to stabilize Iraq and Afghanistan.

In a dramatic departure from the Pentagon's conventional wisdom, the "warrior-intellectuals" aggressively promoted the use of tactical counterinsurgency and eventually succeeded in establishing it as doctrine. Preceding his service in Iraq and during his term as Combined Arms Center commander, Petraeus (with U.S. Marine Corps lieutenant general James Mattis) oversaw the preparation of the U.S. Army and Marine Corps *Counterinsurgency Field Manual*, the first of its kind in more than twenty years.[2] The importance of culture was emphasized throughout the manual: the terms "culture" and "cultural" together appear 178 times in the manual, and among the main contributors was Montgomery McFate, who holds a doctorate in cultural anthropology.[3]

The media showered the soldier-scholars with lavish praise. *Time* magazine described the manual as "radical," "revolutionary," and "Zen tinged."[4] University of Chicago Press reprinted the manual with an introduction by Sarah Sewall of Harvard University's Kennedy School of Government. This edition was clad in an olive-drab book jacket and received adulatory praise from the *New York Times* ("landmark," "paradigm-shattering"), the *Los Angeles Times* ("nifty"), and the *Chicago Tribune* ("probably the most important piece of doctrine written in the past 20 years").[5] Petreaus's right-hand man, John Nagl, appeared on Comedy Central's *The Daily Show* to promote the book.[6]

A few anthropologists embraced the Petraeus doctrine. For example, the U.S. Army War College's Strategic Studies Institute sponsored a sympathetic study by Sheila Miyoshi Jager, who wrote, "The post-Rumsfeld Pentagon has advocated a 'gentler' approach, emphasizing cultural knowledge and ethnographic intelligence. . . . This 'cultural turn' within DoD highlights efforts to understand adversary societies and to recruit 'practitioners' of culture, notably anthropologists, to help in the war effort. . . . [T]he recent focus on cultural knowledge in counterinsurgency operations and tactics is a welcome development."[7]

Robert Gates, who served as secretary of defense after Donald Rumsfeld, also became a strong supporter. In a November 2007 speech, Gates (who has a Ph.D. in history from Georgetown University) outlined the need for "soft power . . . beyond the guns and steel of the military" and favorably mentioned the work of anthropologists serving as military advisers in Afghanistan.[8] Several months later, speaking to the Association of American Universities, Gates proposed the Minerva Consortium, an initiative designed to provide $100 million for university-based social science research to help meet the Pentagon's needs. As evidence of fruitful collaboration between the Pentagon and social scientists, he portrayed the U.S. Army's experimental Human Terrain System program in glowing terms, describing it as key to long term success in Iraq, Afghanistan, and beyond.[9]

Petraeus's military career has ended, and nearly all of his associates have either retired, gone to work for the private sector, or secured academic positions. Even counterinsurgency has fallen out of fashion as the Obama administration's military strategy has shifted from "long-term nation-building with large military footprints" to "low-cost, small-footprint approaches."[10] Yet the Pentagon's interest in cultural knowledge has survived. A telling example is the following statement by General Roy Odierno, U.S. Army chief of staff: "Nothing is as important to your long term success as understanding the prevailing culture and values. Before the most recent set of conflicts, it was generally believed that cultural awareness was only required in select Army units, such as Special Forces or Civil Affairs. Recent history has made clear that we need expanded levels of cultural and regional awareness in *all* Army units."[11]

The reasons for the Pentagon's interest in culture may be clear enough, but what does culture mean to military officials or to soldiers on the ground? What roles are social scientists playing as the military attempts to integrate cultural knowledge into its training? And finally, what does the future hold for this significant transformation of war?

To get a fuller picture of the military's efforts to incorporate cultural awareness and ethnographic knowledge, this chapter examines four overlapping phenomena that are indicative of the Pentagon's cultural turn: (1) the rapid development of linguistic and cultural tools and technologies for use in Afghanistan and Iraq; (2) the Human Terrain System program, designed to embed social scientists in combat brigades; (3) initiatives aimed at the creation of modeling, simulation, and forecasting software that uses social science data and theories;

and (4) the Pentagon's burgeoning network of in-house anthropologists, culture training centers, and social science research programs.

From World War II Handbooks to Digital Handheld Translators

By now it is well known that the U.S. Army and Marine Corps did not have a unified cultural training program or policy when American troops invaded Afghanistan in 2001 and Iraq in 2003. Consequently, officers and combat troops arrived largely unprepared for the occupations that followed the fall of the Taliban and Saddam Hussein's government, respectively.[12] A handful of observers had expressed concern about this deficiency during the 1990s,[13] but they had been largely ignored.

It is puzzling that Pentagon planners would have overlooked culture. During World War II, the U.S. armed forces relied heavily on anthropological insights. Ruth Benedict famously helped military planners prepare for the occupation and reconstruction of postwar Japan. Others helped the Office of Naval Operations write civil affairs handbooks for a better understanding of Marshall Islanders, Caroline Islanders, and others who had been living under Japanese colonial rule but were now subject to American occupation during the postwar period. Still other anthropologists were employed as analysts by the Office of Strategic Services (the precursor to the CIA) or the War Relocation Authority, a government agency tasked with interning more than a hundred thousand Japanese and Japanese Americans.[14]

In 1943 the U.S. Army published a small booklet titled *Instructions for American Servicemen in Iraq during World War II*, which among other things included simple historical, religious, geographical, and cultural information for soldiers. Perhaps the most interesting parts of the book are the tidbits of advice it offers: "You aren't going to Iraq to change the Iraqis. Just the opposite. We are fighting this war to preserve the principle of 'live and let live.'" "Don't stare at anyone who is praying, above all do not make fun of him. Respect his religion and he will respect yours." "Talk Arabic if you can to the people. No matter how badly you do it, they will like it."[15]

During the 1950s and 1960s, U.S. military and intelligence agencies found multiple uses for cultural knowledge and anthropological research—far too many to be described in this chapter. David Price's research reveals that, during this period, some anthropologists willingly complied with the needs of the national security state, while others were unwilling participants in these processes.[16] The U.S. Army created the Special Operations Research Office in the late 1950s, and this eventually moved to the Center for Research on Social Systems at American University, which also housed the Counterinsurgency Information Analysis Center. These three entities successfully recruited social scientists—including anthropologists—who could provide analysis and regionally specific knowledge for military use.

Other projects focused on cultural education and training. The Personal Response program, funded by the U.S. Army and Navy, was a language and cultural training program for those deployed as part of the Marine Corps Combined Action Platoons, essentially counterinsurgency teams made up of Marines and anticommunist Vietnamese villagers.[17] At about the same time, the Army Research Office contracted the American Institutes for Research to develop the Troop-Community Relations program, an initiative aimed at improving relations between U.S. Army troops and their Korean hosts. The program was eventually expanded to Thailand.[18] It is worth noting that in both the Personal Response and Troop-Community Relations programs, the culture concept was largely derived from the fields of psychology and communication studies—what would be known today as cross-cultural communication. Also worth noting is that none of these programs' architects appear to have been trained anthropologists (as was the case among some staff of the Special Operations Research Office, Center for Research on Social Systems, and Counterinsurgency Information Analysis Center).[19]

That is not to say that anthropologists did not participate in the Vietnam War effort. Their involvement in military programs during this period tended to be more analytic, strategic, and tactical than educational. In 1964, the Special Operations Research Office initiated the ill-fated Project Camelot, an army program designed to uncover the causes of insurgencies and determine steps that governments might take to counter such movements—including counterinsurgency operations. When Hugo Nutini, an anthropologist who had participated in the conceptual phase of the project, offered to recruit Chilean social scientists he aroused suspicion. As word leaked out about the program and its objectives, Nutini and other participants were criticized harshly by the media and the academy. After congressional hearings on Project Camelot were held in 1965, defense secretary Robert McNamara abruptly terminated the program.[20]

During the same decade, a group of American and Australian anthropologists conducted research in support of U.S. counterinsurgency operations in Thailand. News about this work went public after a University of California at Los Angeles graduate student stole documents from Michael Moerman, a participating anthropologist. Because this work was carried out without public knowledge and with potentially devastating results for native peoples, there was an outcry that eventually resulted in deep divisions within the American Anthropological Association and eventually the adoption of the association's first code of ethics.[21] What is notable about both Project Camelot and the so-called Thai Affair is that cultural knowledge was clearly being mobilized as a weapon. Laura Nader has persuasively argued that the impact of such Cold War programs on anthropology was complex and often resulted in secrecy and silence. Some anthropologists reacted by launching sharp critiques of the national security state and by analyzing powerful public and private institutions. Others responded by reorienting the discipline toward trivial, esoteric topics and writing in an inaccessible style filled with academic jargon.[22]

Perhaps, then, it is not surprising that from the 1970s to the early 2000s, cultural considerations largely disappeared from the Pentagon's agenda. When President George W. Bush declared war on terror in 2001, it quickly became evident that the Pentagon's earlier experiments with culture had long been forgotten.

The military's lack of cultural preparation in Iraq and Afghanistan was initially dealt with in a clumsy manner. To address the immediate problem of an insufficient number of human translators, the Pentagon ordered thousands of handheld translation devices from companies such as VoxTec and Integrated Wave Technologies to help troops communicate with Pashto, Urdu, and Arabic speakers.[23] The problem, however, was that machines like the Phraselator (VoxTec's device) "only [came] close to working as long as speakers confine[d] themselves to one sentence of speech at a time and ha[d] the right accent."[24] One can also imagine how such devices might miss the subtleties of language (metaphors, irony, slang, and emotion, to name but a few) and how they might be impractical at those moments in which linguistic understanding is crucial—for example, during moments of confrontation, panic, or violence.

Military agencies introduced another heuristic device—a decidedly low-tech one—in 2003. The Iraq Culture Smart Card was created by the U.S. Marine Corps Intelligence Activity for use by ground troops. The Culture Smart Card was a laminated pamphlet consisting of sixteen panels that could be folded up like a pocket-sized map. Each panel was filled with factoids about various topics such as "Ethnic Groups," "Clothing/Gestures," and "Helpful Words and Phrases." A similarly structured Afghanistan Culture Smart Card was released late in 2004. By 2006, nearly two million cards had been produced.[25]

As with the Phraselator, the idea of pocket-sized culture might sound appealing, but the Culture Smart Cards were not always successful or even useful. Rochelle Davis notes that although some of the panels were "serviceable," other panels misinterpreted meanings or attributed incorrect meanings to things (such as clothes or headscarves). She writes:

> These materials do a disservice to both Iraqis and US servicemen and women themselves. Pedagogically, the presentation fails to make clear why the factoids are important. . . . [T]he vast majority of the 50 soldiers and Marines interviewed for this article assessed their formal cultural training as "not useful" because "culture" was described as a fixed reaction or behavior, often a list of dos and don'ts that could be obeyed like orders, rather than a contextual understanding.[26]

By 2005, U.S. casualties were mounting, Iraqi insurgent groups were becoming stronger, and Taliban fighters were regrouping. As Robert Gates and David Petraeus gained influence, counterinsurgency and culture came to be seen as necessary components for battlefield success. An early advocate was Major General Robert Scales, who told the U.S. House Armed Services Committee that "the British Army created a habit of 'seconding' bright officers to various corners of the world so as to immerse them in the cultures of the Empire. . . . At the heart of a cultural-centric approach to future war would be a cadre of global scouts. . . . They should attend graduate schools in disciplines necessary to understand human behavior and cultural anthropology."[27] The groundwork was set for cultural-centric warfare and weaponized ethnography.

Scales would not need to wait long. In 2005, Montgomery McFate and Andrea Jackson published a pilot proposal for a Pentagon "Office of Operational Cultural Knowledge" focused on "human terrain" and consisting of social scientists with "strong connections to the services and combatant commands."[28] This would eventually become the framework for the Human Terrain System program.

Embedded Ethnography

Between July 2005 and August 2006, the U.S. Army assembled an experimental program called the Human Terrain System (HTS). Its building blocks were five-person teams (human terrain teams, or HTTs) assigned to combat brigades in Iraq and Afghanistan and composed of regional studies experts and social scientists, some of whom were armed.[29] According to the program's architects, the embedded social scientists would serve as cultural advisers to brigade commanders. A sympathetic reporter described the teams as "a 'graduate-level counterinsurgency' unit . . . [created] to fine-tune aid and to undermine the intimidating grip of militants."[30] Within five years, HTS expanded from five teams to thirty and employed more than five hundred people. Since HTS began, the U.S. government has spent $726 million on it, making it the most expensive social science program in history.[31]

The original proposal for HTS laid out ambitious goals, including "on-the-ground ethnographic research" on the Middle East, Central Asia, and elsewhere; "predeployment and advanced cultural training . . . [and] computer-based training on society and culture"; "sociocultural studies of areas of interest"; "cultural advisers for planning and operations to commanders on request"; and a comprehensive "constantly updated database tool for use by operational commanders and planners."[32]

Soon after, Jacob Kipp and colleagues from the U.S. Army's Foreign Military Studies Office at Fort Leavenworth, Kansas, outlined the HTS as aiming to "understand the people among whom our forces operate as well as the cultural characteristics and propensities of the enemies we now fight."[33] Captain Don Smith headed the implementation of HTS from July 2005 to August 2006, and the program was housed in the Training and Doctrine Command (TRADOC) at Fort Leavenworth. Each team was envisioned to consist of a mix of personnel that would include military intelligence officers and cultural experts (preferably anthropologists with graduate degrees).

In early 2007, the British defense firm BAE Systems began posting HTS job announcements on its website. Soon afterward teams began arriving in Afghanistan. Others deployed to Iraq in summer 2007.

As HTS expanded, its proponents insisted that HTTs were "extremely helpful in terms of giving commanders on the ground an understanding of the cultural patterns of interaction, the nuances of how to interact with those cultural groups on the ground"—but in reality, none of the Ph.D.-qualified anthropologists working in HTTs had prior regional knowledge when Montgomery McFate made this claim.[34] Even so, HTS grew rapidly, partly because of the rapid adoption of counterinsurgency doctrine by the Pentagon.

News organizations generally described HTS and its leadership in glowing terms following its implementation. The *New York Times* ran a lengthy cover story about the program in October 2007 and was soon followed by *U.S. News and World Report*, *Christian Science Monitor*, *Time*, and the *Wall Street Journal*, among others.[35] Extensive profiles of Montgomery McFate were published by the *San Francisco Chronicle Magazine* and *Elle*.[36] CNN anchorman Tom Foreman conducted a friendly interview with McFate on the program *This Week at War* without asking any critical questions.[37] At about the same time, various radio shows (mostly affiliated with National Public Radio) broadcast stories about HTS while giving little time for critics to speak out against the program. PBS talk show host Charlie Rose gave McFate the red-carpet treatment, ending a gushing interview with the question "How come you're so smart?"[38]

The media blitz might have been the result of a carefully orchestrated public relations campaign, of news editors desperately seeking good news from Iraq and Afghanistan, or both. Yet HTS's alleged success was supported by only limited anecdotal evidence. Despite HTS supporters' claims that the program had dramatically reduced U.S. "kinetic operations" (military attacks) in Afghanistan, independent assessments were not conducted until years later. When these finally appeared, they revealed no concrete evidence supporting the claim that HTS contributed to a reduction in kinetic operations.[39]

In the meantime, HTS experienced a series of setbacks. During the summer of 2007, the Network of Concerned Anthropologists was formed to oppose anthropological involvement in counterinsurgency work. (I was one of the cofounders.) More than a thousand people joined the network within months and pledged to refrain from counterinsurgency support. The American Anthropological Association executive board issued a statement in November 2007 expressing disapproval of HTS.[40] In May 2008, the Society for Applied Anthropology expressed "grave concerns about the potential harmful use of social science knowledge and skills in the HTS project."[41] In December 2009, the American Anthropological Association went even further when it publicly released a report conducted by its ad hoc Commission on the Engagement of Anthropology with US Security and Intelligence Communities, stating that

> when ethnographic investigation is determined by military missions, not subject to external review, where data collection occurs in the context of war, integrated into the goals of counterinsurgency, and in a potentially coercive environment—all characteristic factors of the HTS concept and its application—it can no longer be considered a legitimate professional exercise of anthropology.[42]

Relatively few anthropologists responded to HTS's recruitment efforts. As of April 2009, only 11 HTS employees out of 417—that is, less than 3 percent—held graduate degrees in anthropology.[43]

In the meantime, HTS personnel suffered a series of tragic events. On May 7, 2008, Michael Bhatia—a political scientist and HTS social scientist—was killed in a roadside bomb attack in Afghanistan. Several weeks later, another HTS employee, political science graduate student Nicole Suveges, was killed in a bomb attack in Sadr City, Baghdad. In November 2008, yet another HTS team member, Paula Loyd, was doused with kerosene and set alight by an Afghan man. She died from her injuries on January 7, 2009.[44]

By the end of 2009, HTS was coming under fire from seasoned military officials. Marine Corps major Ben Connable complained that the program was siphoning away resources and staff—notably foreign affairs and civil service officers and intelligence personnel. He asked, "Why is it necessary to create a separate program, costing (at a minimum) tens of millions of dollars?"[45]

Another critic, U.S. Army colonel Gian Gentile, challenged the claims of HTS leadership. He noted, "The effectiveness of the HTTs was dubious at best . . . even if the HTT program was able to match specialists to the areas in Iraq and Afghanistan in which they were expected to operate, it still would not have made any kind of significant difference in the outcome of these wars."[46]

Despite these concerns, HTS made the transition from proof of concept to program of record (i.e., a permanent Army program) in 2010. Montgomery McFate and Steve Fondacaro largely ignored critics and instead attributed HTS's "challenges" to the program's rapid growth. In their words, HTS's growing pains were evidence of its "catastrophic success."[47]

But HTS might have been more accurately called a successful catastrophe. McFate and Fondacaro left the embattled program in 2010, but even under new leadership HTS did not escape controversy. If anything, the criticism intensified. Beginning in early 2013, a series of reports by Tom Vanden Brook of *USA Today* revealed endemic and continuing patterns of waste and abuse plaguing HTS. The reports indicated that HTT members had been encouraged to inflate their time sheets in order to maximize pay, that racial and sexual harassment had run rampant for years in the program, and that HTS had received little if any oversight since its creation. The reports also revealed that standard team members' annual salaries ranged from $224,000 to $280,000. By 2012 at least some reports implied that funding for HTS might be cut as the result of the Pentagon's shrinking budgets, but the program managed to survive until September 2014, at which time it was quietly terminated.[48]

In a nuanced and carefully researched analysis of HTS, investigative journalist Vanessa Gezari offered the following summary of the program:

> [HTS] had everything to do with the contradiction between the United States' self-image as a benevolent superpower and the realities of war and the economy that drives it. . . . In fact,

> the program was a giant cultural metaphor, a cosmic expression of the national zeitgeist: American exceptionalism tempered by the political correctness of a postcolonial, globalized age and driven by the ravenous hunger of defense contractors for profit. If you could have found a way to project on a big screen the nation's mixed feelings about its role as the sole superpower in a post–Cold War world, this was what it would have looked like.[49]

We will soon revisit the theme of the American public's "mixed feelings" about being the world's "sole superpower." Now let us turn to another group of emergent Pentagon projects that make use of cultural and ethnographic data: modeling, simulation, and forecasting programs.

Cybernetic Crystal Balls and Virtual Battlefields

Over the past few years, U.S. military and intelligence agencies have spent hundreds of millions of dollars developing computer programs to predict hot spots ranging from organized protest marches to riots to full-blown insurgent attacks.[50] Millions more are being dedicated to "analyzing the emotional temperature of postings on Facebook and Twitter, or the telephone traffic between groups of villages" in the Middle East, Central Asia, or other areas experiencing political tension or civil conflict.[51] Much of the research has been funded by the Human Social Culture Behavior Modeling Program, which was created in 2008 by the Office of the Secretary of Defense as a six-year research initiative.[52] Those spearheading such efforts are a rapidly growing group of engineers, mathematicians, and computer scientists seeking to synthesize a wide range of data—including ethnographic data—into their programs. Many of the programs use rational choice and social movement theories as the basis for modeling.

Among the researchers is Barry Silverman, who has developed computerized behavior modeling software designed to uncover the hidden motivations of terrorists and their networks.[53] These programs, funded by the Pentagon's Defense Modeling and Simulation Coordination Office, integrate "more than 100 models and theories from anthropology, psychology, and political science, combined with empirical data taken from medical and social science field research, surveys, and experiments."[54] Simulations represent "physical stressors such as ambient temperature, hunger, and drug use; resources such as time, money, and skills; attitudes such as moral outlook, religious feelings, and political affiliations; and personality dispositions such as response to time pressure, workload, and anxiety."[55] Among Silverman's best-known programs is NonKin Village, an interactive training game that resembles the popular video game SimCity. NonKin Village uses models based on cultural data from specific societies to create avatars who interact with the player. The goal is for players to experience foreign cultures and demonstrate sensitivity to local customs, practices, and values.[56]

Gary Ackerman, director of the Center for Terrorism and Intelligence Studies (a division of Akribis Group, a security company based in San Jose, California), provides a vivid description of the way modeling and simulation programs work: "There are tools where they build a world in a bottle. They put down every single mosque, river, camel, and school in, say, Saudi Arabia. Then they have millions of software agents who each have desires, grievances, all these different variables. They go about their little lives and then you ask a question: What if we build a McDonald's in Mecca? Does this lead to more people joining terrorist groups or not?"[57]

Another wartime simulation project includes Purdue University's Synthetic Environment for Analysis and Simulation, which "gobble[s] up breaking news, census data, economic indicators, and climatic events in the real world, along with proprietary information such as military intelligence."[58] Purdue's Iraq and Afghanistan computer models each have approximately five million individual nodes representing buildings, infrastructural features, mosques, and people.

Other researchers are creating three-dimensional computerized avatars designed to improve the cultural competency of American soldiers. Computer scientists Glenn Taylor (of Soar Technology) and Ed Sims (of Vcom3D) have developed interactive characters that model the physical and cognitive behaviors of people from different societies. Their efforts focus on modeling the behavior of Arabs, and to this end Taylor and a colleague have identified more than two hundred physical gestures from interviews with Baghdad residents. Their goal is the creation of a training tool that relies on humanoid avatars:

"[Our software system] has developed a library of culture-specific avatars, gestures and expressions that can be invoked on demand. These libraries consist of over 60 culturally diverse virtual human models as well as 40 facial expressions and 200 gestures, and can automatically lip-sync to over 22 mouth shapes that map to over 100 speech sounds of the International Phonetic Alphabet."[59]

This software also attempts to integrate "cultural cognitive architecture" that uses "knowledge learned by living in that culture: knowledge about the correct ways to interact (e.g., norms), about what is important in the culture (e.g., values), etc."[60] By creating virtual plug-and-play Iraqis, the researchers hope to provide a low-cost, low-risk alternative to cultural immersion.

Alelo, a Los Angeles–based company, creates a number of products used by the Pentagon, including the Virtual Cultural Awareness Trainer and also the Tactical Language and Culture series. These are three-dimensional interactive computer games—virtual role-playing games—designed to teach military personnel about Iraqi and Afghan languages and values. The player views an animated version of an Iraqi or Afghan village or neighborhood from a first-person point of view and converses and interacts with people to understand the society and then to work with the community to affect the outcome of military missions. Virtual villagers form opinions based on the way the player interacts with them, while the player gathers data about characters from his or her observation of verbal and nonverbal cues.[61]

Still others are involved in the creation of forecasting programs designed to predict future events based on the analysis of multiple streams of data, including sociocultural, political, economic, medical, and geographical. For example, Mark Maybury, an artificial intelligence specialist who served as the U.S. Air Force's chief scientist from 2010 to 2013, is attempting to develop a "Social Radar" capable of "seeing into the hearts and minds of the people."[62] The idea is to use "sociometrics," including "Facebook timelines, political polls, spy drone feeds, relief workers' reports, and infectious disease alerts" to give early warnings.

Lockheed Martin's World-Wide Integrated Crisis Early Warning System (W-ICEWS) may well be the biggest forecasting initiative of all. DARPA (Defense Advanced Research Projects Agency) has reportedly spent $38 million on W-ICEWS, which consists of four components: iData, a massive database holding approximately thirty million cross-indexed news stories from six thousand publications; iTrace, a tool that uses open-source documents and text analytics to monitor political activity on a global scale; iCast, a software program that provides six-month forecasts for destabilizing events in 167 countries; and iSent, a means of tracking regional sentiment by using social media as intelligence data.[63] According to Noah Shachtman, W-ICEWS is designed to combine macrostructural and real-time event data with the perspectives of regional experts and then "ground predictions in social science theory."[64]

It appears that the Pentagon has decided to cast a wide net in an effort to develop programs for predicting the future. Other modeling, simulation, and forecasting programs are under development—far too many to review in this chapter. Despite these programs having generally done a poor job of predicting the future, it seems that computerized algorithmic tools make sense within the cultural framework of a military that has long had a preference for high-tech, high-cost solutions.[65]

Among the most vocal critics of forecasting programs are people within the military establishment. Lieutenant Colonel John Nagl told *Wired* magazine, "Wait a minute, you can't tell me who's going to win a football game. And now you're going to replicate free will?"[66] When asked about the engineers and computer scientists developing such programs, retired lieutenant general Paul Van Riper told *Science* magazine, "They are smoking something they shouldn't be. Only those who don't know how the real world works will be suckers for this stuff."[67]

These examples illustrate a growing field in which "culture," "tribes," "moral outlook," and other constructs appear as independent variables in complex mathematical formulas that view *Homo sapiens* as a machine governed by a calculus of the psyche—an algorithm of the soul.

Perhaps it is for this reason that the Pentagon has also committed itself to the creation of other virtual worlds, not just computerized role playing games. For example, at the National Training Center (NTC) in Fort Irwin, California, the Army has constructed several mock villages from scratch. In these villages, located in the middle of the Mojave Desert, real-life Iraqi and Afghan Americans are hired for twenty-eight-day stints in which they play assigned roles of insurgents, allies, and neutral villagers while U.S. combat units practice counterinsurgency techniques in a virtual battlefield. NTC employs Hollywood producers, directors, actors, and special-effects technicians to help develop scripts and scenarios, train role players, and create explosive special effects—all to replicate the sights, sounds, and smells of occupied Iraq.[68] The surreal war theater blurs the line between fantasy and reality.

Is it possible that in the not-too-distant future, military planners will use ethnographic data, social science theories, and predictive modeling for preemptive targeting of statistically probable insurgents in Syria, Iraq, Afghanistan, Pakistan, Yemen, or other countries deemed to be terrorist havens? Perhaps we are not so far away from the dystopian visions of science fiction writer Philip K. Dick. In his story *The Minority Report*, Dick describes a society in which three babbling mutants foresee crimes before they occur.[69] The mutants produce data that are channeled into powerful computers designed to help police prevent crimes, but the story's protagonist learns that mutants and computers can misjudge future events.

In this context it is worth revisiting the cautionary words of C. Wright Mills, perhaps more applicable today than when he wrote them more than a half century ago:

> To talk so glibly as many do about prediction and control is to assume the perspective of the bureaucrat to whom, as Marx once remarked, the world is an object to be manipulated. . . . But we, as social scientists, may not assume that we are dealing with objects that are so highly manipulable, and we may not assume that among men we are enlightened despots. . . . No historical society is constructed within a frame as rigid as that enclosing my hypothetical army division. Nor are social scientists—let us be grateful—generals of history.[70]

Seen through an anthropological lens, twenty-first-century modeling, simulation, and forecasting technologies appear not so much as scientifically based tools for confronting knowable, predictable phenomena but rather as amulets for dealing with dangerous, unknowable events by means of sacred formulae (algorithms). These computer programs, software packages, and cybernetic tools might even be seen as forms of magic, which Pentagon planners and their corporate contractors employ in unpredictable, risky situations—popular uprisings, armed rebellions, peaceful revolutions, and demands for radical democratic change—over which they have little control.

The Military's In-House Culture Experts

Over the past decade, the Pentagon and its military branches have rapidly assembled a network of specialists and centers dedicated to cultural knowledge training and acquisition. These efforts to develop in-house culture experts emerged in the wake of U.S. counterinsurgency operations in Iraq and Afghanistan. Once it became clear that few academic anthropologists would respond to DoD recruiting efforts, the Pentagon began building its own "parallel universe" of specialists.[71] The goals were to create a group able to train officers and other military personnel about language and culture, to exploit the knowledge of regional specialists or "subject matter experts" able to inform military operations, and to develop cultural handbooks, guidelines, and tools for use by deployed troops and military planners.[72]

Changes in the Marine Corps are a striking example of the new approach. In 2005, its Center for Advanced Operation Cultural Learning opened in Quantico, Virginia. The center prepares Marines "to work with local populations and partner militaries," to apply "operational culture" for military initiatives, and to deliver "mission-oriented culture and language" to deploying units.[73] Some of the center's efforts include creating and compiling guidebooks such as the *Cultural Intelligence Indicators Guide*, described by Stephen Fallon as "a 'do-it-yourself' guide for marines to track their cultural lessons and intelligence gathering." "This novel approach treats every man as a data collector, steering marines by providing suggestions on areas to inquire and investigate further during their deployment."[74] Other guidebooks include *Operational Pashtunwali* and *Operational Culture for Deployment: Afghanistan*.[75]

From an anthropological perspective, it is curious that the center has an obsessive focus with a specific kind of culture: *operational* culture. For example, a leading figure there is Paula Holmes-Eber, whose title is professor of operational culture at Marine Corps University and who coauthored the book *Operational Culture for the Warfighter.* The book, which is essentially an introductory cultural anthropology textbook tailored for Marines, defines operational culture as "those aspects of culture that influence the outcome of a military operation; conversely, the military actions that influence the culture of an area of operations."[76] The book defines the war fighter as a "culture operator" ("who engages in military functions at the tactical, operational, and strategic level within his AO [area of operations] through continually re-reading the changing cultural and human aspects of the battlespace as they impact military operations") and offers critical comments on the "human terrain" concept.[77]

In 2006, the Air Force created the Culture and Language Center, located at Maxwell Air Force Base in Montgomery, Alabama. The Culture and Language Center was designed to introduce cross-cultural awareness across Air University schools, including its non-commissioned officer academies, the Air War College, and its community colleges. In an interview, Bruce Murphy (then Air University's chief academic officer), contrasted the Culture and Language Center with embedded anthropology: "The idea is rather than finding anthropologists and co-locating them with military units . . . our focus is to get all of our airmen, all of our students, to be able to be sensitive to these cross-cultural concepts and be able to use them wherever they're deployed."[78] Anthropologists currently affiliated with the center include Dan Henk, Brian Selmeski, Patricia Fogarty, and Angelle Khachadoorian.

In early 2006, the Army created the TRADOC Culture Center at Fort Huachuca, Arizona, home of the Army Intelligence Center. The Culture Center offers training classes and publishes Smart Books, which provide background information about the peoples of Afghanistan, Pakistan, and Yemen. The center also hosts an annual Culture Summit, in which friendly delegates from different countries come together to discuss culture training. Occasionally, such events reveal the strictly utilitarian view of culture held by some military officials, as when Major General David Hogg proposed that the military use "culture as a weapon system."[79]

The TRADOC Culture Center is perhaps the most outwardly focused of the new culture training centers and even has a Facebook page with short informational videos resembling TED Talks. One video features Colonel John Bird, the Culture Center's director, describing culture as an essential part of "adaptive leadership." The well-rehearsed presentation, which includes a brief exchange between Bird and members of the audience (most of whom wear camouflage fatigues), illustrates the Army's notion of culture as a tool for military operations:

> BIRD: Why do you think culture training might be relevant?
>
> AUDIENCE MEMBER 1: It's relevant, sir, because it allows leaders to better understand the environment they'll be operating in.
>
> AUDIENCE MEMBER 2: It helps us to connect with the population.
>
> AUDIENCE MEMBER 3: We need to know what the normal is.
>
> AUDIENCE MEMBER 4: And it can help us build rapport with the host nation security forces.
>
> BIRD: All excellent responses. I want you to know that this—this thing, culture, this isn't about big group hug; this isn't about winning hearts and minds. This is about winning respect. It's about enabling you to more effectively accomplish your mission.... Don't be afraid of this thing. Understand it. Embrace it. Leverage it.[80]

It appears that at least some of TRADOC Culture Center's training responsibilities are being subcontracted to firms, including General Dynamics.[81] Similarly, the Center for Advanced Operation Cultural Learning appears to subcontract at least some of its curriculum development to companies such as ProSol Associates.

The Navy has also established a culture training school, the Center for Language, Regional Expertise, and Culture, created in 2006 in Pensacola, Florida. According to a recent assessment report, much of the center's efforts have been focused on cultural orientation training coordinated by Mobile Training Teams, "semiautomated" group presentations, and individual instruction, including "e-Learning."[82]

Cultural knowledge for the military has become a cottage industry stretching beyond the DoD, which should not be surprising given the Pentagon's extensive influence. For example, at Norwich University, the oldest private military college in the United States, the Institute for the Study of Culture and Language was created in 2013.

Its director, cultural anthropologist Robert R. Greene Sands, previously held a faculty position at the Air Force Culture and Language Center. The institute's private sector defense companies, such as Rababy Associates, Glevum Associates, Silverback 7, and SCIA (among many others), have also expanded their range of services to include "Face-to-Face Research and Analysis," "human terrain analysis," and "culture research/culture training." Nonprofit research organizations such as Rand have also jumped on the culture bandwagon by sponsoring anthropological and ethnographic research for military consumption in recent years.

For some, the growth of cultural training programs and the Pentagon's in-house anthropology might be seen as welcome developments. After all, who can be against the idea of increasing cultural awareness and competency, whether it is among elite policy makers, military personnel, or gas station attendants?

However, from another perspective, the military's deployment of culture is troubling. Paula Holmes-Eber's recent work reveals its inherent contradictions. She notes that when it was finally implemented, the Pentagon's "new culture policy became 'Marinized'—becoming steeped long enough in Marine Corps ways of doing things that it 'looks, smells, and tastes Marine.'" Her observations raise an important question: What happens to the culture concept (or to empathy for the Other, for that matter) once filtered through the Marine Corps' masculine "cultural ideals of toughness, endurance, and invulnerability to pain"? She reports that, ultimately, cultural knowledge was translated simplistically into "familiar military ways of analyzing information through graphics, PowerPoint presentations, and computerized simulation programs."[83]

Part of the problem is a notion held by many of the Pentagon's in-house anthropologists: that they will successfully "anthropologize the military" once they have spread the gospel of culture.[84]

Although these stated aims of the Pentagon's in-house anthropologists sound laudable and even heroic, they are naive at best and at worst disingenuous. Historical precedents can tell us a great deal. U.S. military and intelligence agencies consistently supported anthropological work during the first two decades of the Cold War, selectively choosing to use research that supported its aims while ignoring that which did not. In a nuanced review of the anthropologist (and Rand counterinsurgency expert) Gerald Hickey's 1960s work in Vietnam during the height of the war, David Price notes, "There is no reason for contemporary anthropologists to not learn from his experiences: some might claim the moral of Hickey's story is that we must work harder to make the military understand what anthropology has to offer, but such interpretations ignore the importance of institutional culture and the possibility of larger contingencies governing the use of military knowledge. . . . [M]otivations can have little impact on outcomes."[85] Hickey is one of a long list of anthropologists who had good intentions yet were ultimately unsuccessful in their efforts to transform military policies and institutional culture.

Some might argue that the Pentagon's training programs can be interpreted as feeble attempts to apply a cultural veneer over the brutal ugliness of war. The words of those who have experienced war from the front lines speak volumes about the contradictions of militarizing culture. Among the most articulate and moving accounts is that of former Marine infantryman Tyler E. Boudreau.[86] In the first few chapters of his book *Packing Inferno*, Boudreau describes how he initially viewed the Marines' mission after reading the anthropologically informed *Counterinsurgency Field Manual*:

> We would do it right. We were going to help the Iraqi people rebuild their country. We were going to get to know the local leaders, establish rapport, fund new projects, bring back electricity and water, construct schools and medical clinics, build new soccer fields. . . . We were going to win their hearts and minds, and every once in a while, we were going to get in a fight with the bad men who didn't want to see progress in their country and who couldn't fathom the peace. General Mattis said to us: You've got to know their culture, so you don't inadvertently aggravate the people and invigorate the insurgents.[87]

But within days after arriving in Iraq, Boudreau finds himself descending into a world that he can only compare to Dante's inferno. Once fighting begins, cultural niceties evaporate, and the Marine Corps' "killer culture" prevails. Reflecting on his journey, Boudreau uncovers the irreconcilable contradictions between cultural sensitivity and the process of desensitization that is a routine part of Marine training, beginning in boot camp:

> They taught us martial arts—the axe kick—swift driving heel into the enemy's face as he lay immobilized on the ground. They promised if we did it right, we'd leave his skull a broken mess. They taught us how to sneak up behind the enemy and dig a combat knife up under his rib cage, how to gouge out his eyeballs, how to cut off the supply of blood to his head, or air to his lungs, and how to rip out his testicles. We learned how to butt-stoke him in the face, thrust our bayonets into his flesh, and we practiced on hanging dummies. These weren't the inanimate tires stacks of the old days. These dummies were dressed in uniforms, and they were heavy and dense like a man, and we screamed "kill" as our blades punctured their skins. Most of all, we learned to look upon the act of killing with nonchalance. . . . Killing is the culture the soldier has entered.[88]

One cannot help but wonder whether the extreme cognitive dissonance induced by boot camp desensitization processes, on the one hand, and cultural training programs, on the other, might be at least partly responsible for the daily agonies faced by Boudreau and countless others (including some who are now in our university classrooms):

> The Marine Corps made a schizophrenic out of me, figuratively speaking. . . . My mind changed. It split, two, and three, and four ways over. Now there are voices coming at me from all sides. There's no coherence to them, and no clear distinction either. There's the angry voice, and there's the broken-hearted one. There is the tender me and the savage. And of course, there is the Marine. There will always be the Marine, standing tall inside me, speaking smartly about values and patriotism. Then there's all the rest of me, the part of me that was left over when I left the Corps. He has no name, no identity, or credentials, or skills. He has no title or rank. He has no cause. He is just me in the wake of battle.[89]

Conclusion: Culture, Counterinsurgency, and Common Sense

Those responding to the Pentagon's call for cultural knowledge frequently overlook an important point: counterinsurgency itself is a cultural system, replete with its own logic, structures, symbols, values, and unquestioned assumptions. For example, recent counterinsurgency doctrine assumes that culture can be used as a way of reducing "native perceptions of the United States as an occupying power"[90]—or to put it more bluntly, to "leverage social structure and hegemonic narratives so that the occupied will internalize their own captivity as 'freedom.'"[91]

Similarly, the Pentagon's in-house anthropologists have eagerly provided the technical, methodological, and educational expertise sought by the DoD, but few if any have been willing or able to contextualize and analyze the broader significance of their work. Even a cursory historical comparison reveals problems with militarized anthropology:

> British military and civilian officers did not need to hire anthropologists, because they themselves were the experts who would spend their entire careers in one geographical area, in the process learning the languages and customs of the people they governed. Their families accompanied them, and they set up homes and raised families [in the colonies]. . . . They were there for the duration, in other words, as was the government they represented, and the result was that they had no choice but to get it right. Before accepting that it is necessarily the right direction for our military to go, we should pause a bit and consider what embracing this [counterinsurgency] doctrine really means. . . . [T]he British embraced the role of an imperial power and to date, the Americans have not. In truth, the Americans do not know what their role is or should be.[92]

Or in other words, can the U.S. military ever be "anthropologized"—or become a colonial service agency or a humanitarian aid organization—given that the United States is an empire in denial?

Stepping back for a moment from the utilitarian, uncritically applied uses and abuses of cultural knowledge, it is worth considering how anthropological insights might lead to deeper understandings and more incisive questions: Why did DoD officials largely ignore culturally informed perspectives for more than four decades, even during periods of budgetary growth? When Pentagon officials finally did incorporate cultural knowledge, why were they only interested in operational forms of culture—pocket-sized culture, culture as a force multiplier, culture as an algorithm, culture as a weapon?

Stepping back further, we might ask even broader questions: How can we ensure that wars be less likely to occur rather than more humane once they begin?[93] "What was it about decision-making processes at the top of the National Security State that allowed people . . . to think that they would be welcomed with flowers" in Iraq and Afghanistan?[94] And why has the ideology of American exceptionalism functioned reliably as a controlling process for so long in our society? The words of Franz Boas, the father of modern American anthropology, are as insightful today as they were nearly a century ago: "The American, on the whole, is inclined to consider American standards of thought and action as absolute standards; and the more idealistic his nature, the more strongly he wished to 'raise' everyone to his own standards. . . . Therefore, he is inclined to assume the role of a dispenser of happiness to mankind. . . . I have always been of the opinion that we have no right to impose our ideals upon other nations, no matter how strange it may seem to us that they enjoy the kind of life they lead."[95]

Militarized cultural knowledge—like the Pentagon's recent experiments with counterinsurgency—is being deployed as much for domestic use as it is for direct application by commanders in faraway lands. Images of uniformed anthropology professors lecturing to uniformed officers or combat troops are visually appealing symbols ready-made for American viewing and listening audiences. These images, along with others depicting soldiers sitting on pillows drinking tea with Afghan leaders or distributing soccer balls and sweets to Iraqi children, communicate an unambiguous message to the American public: this is a kinder, gentler, and smarter war. This is a war we can feel good about fighting, the images tell us. But the reality is that military operations, expeditionary missions, and the very notion of "operational culture" are fundamentally at odds with the anthropological approach, which relies on mutual respect, empathy, and reciprocal relations toward those from other societies—as well as a cosmopolitan perspective acknowledging that the American way of life is but one among many. Ultimately, militarized ethnography is doomed to fail because it reproduces and reinforces the assumptions, structures, and practices of the very military institutions that underwrite it.[96]

Notes

1. Thomas Ricks, "Officers with PhDs Advising War Effort," *Washington Post*, February 5, 2007, p. A1. Among the most influential members of Petraeus's inner circle were Lieutenant Colonel David Kilcullen (Ph.D. in politics, University of New South Wales), Lieutenant John Nagl (Ph.D. in history, Oxford's St. Anthony's College), Colonel Michael Meese (Ph.D. in economics, Princeton University), Lieutenant Colonel H. R. McMaster (Ph.D. in history, University of North Carolina at Chapel Hill), and Colonel Peter Mansoor (Ph.D. in history, Ohio State University).

2. U.S. Army and Marine Corps, *Counterinsurgency Field Manual* (Chicago: University of Chicago Press, 2007).

3. Patrick Porter, "Good Anthropology, Bad History," *Parameters* 37 (Summer 2007): 48; Fred Kaplan, *The Insurgents: David Petraeus and the Plot to Change the American Way of War* (New York: Simon and Schuster, 2013), 214.

4. Joe Klein, "Good General, Bad Mission," *Time*, January 12, 2007, available at http://content.time.com/time/nation/article/0,8599,1587186,00.html.

5. Samantha Power, "Our War on Terror," *New York Times*, July 29, 2007, available at http://www.nytimes.com/2007/07/29/books/review/Power-t.html; David L. Ulin, "What Is Counter-insurgency Good For?" *Los Angeles Times*, July 13, 2007, available at http://latimesblogs.latimes.com/jacketcopy/2007/07/what-is-counter.html; Robert Bateman, "U. of C. Publishes Military Field Manual," *Chicago Tribune*, September 8, 2007, available at http://articles.chicagotribune.com/2007-09-08/entertainment/0709060402_1_doctrine-manual-war-next-week.

6. "John Nagl," *The Daily Show*, August 23, 2007, available at http://thedailyshow.cc.com/videos/dt3sbh/lt--col--john-nagl.

7. Sheila Miyoshi Jager, *On the Uses of Cultural Knowledge* (Carlisle, PA: U.S. Army War College, 2007), 3, 23.

8. Robert Gates, "Landon Lecture," speech given at Kansas State University, November 26, 2007, available at https://www.k-state.edu/landon/speakers/robert-gates/transcript.html.

9. Whitney Kassel, "The Army Needs Anthropologists," *Foreign Policy*, July 28, 2015, available at http://foreignpolicy.com/2015/07/28/the-army-needs-anthropologists-iraq-afghanistan-human-terrain.

10. U.S. Department of Defense, *Sustaining US Global Leadership: Priorities for 21st Century Defense* (Washington, DC: U.S. Government Printing Office, 2012).

11. Ray Odierno, "Regionally Aligned Forces: A New Model for Building Partnerships," *Army Live*, March 22, 2012, available at http://armylive.dodlive.mil/index.php/2012/03/aligned-forces (emphasis in original).

12. Rochelle Davis, "Culture as a Weapon," *Middle East Research and Information Project* 255 (Summer 2010): 297.

13. See, for example, Anthony C. Zinni, "Non-traditional Military Missions: Their Nature, and the Need for Cultural Awareness and Flexible Thinking," in *Capital "W" War: A Case for Strategic Principles of War*, ed. Joe Strange (Quantico, VA: Marine Corps War College, 1998), 247–283; B. C. Lindberg, "Culture: A Neglected Aspect of War" (unpublished thesis, U.S. Marine Corps Staff College, 1996), available at http://www.au.af.mil/au/awc/awcgate/usmc/lindberg.htm.

14. David H. Price, *Anthropological Intelligence: The Deployment and Neglect of American Anthropology in the Second World War* (Durham, NC: Duke University Press, 2008).

15. U.S. Army (Special Service Division), *Instructions for American Servicemen in Iraq during World War II* (1943; repr., Chicago: University of Chicago Press, 2007). The lack of elementary cultural knowledge in the American military is illustrated by John Nagl's comment about the booklet: "Gosh, I wish I'd had it when I deployed to Iraq in September 2003." Quoted in Dwight Garner, "'You Win This War by Drinking Tea,'" *New York Times*, August 14, 2007, available at http://artsbeat.blogs.nytimes.com/2007/08/14/you-win-this-war-by-drinking-tea.

16. David Price, *Cold War Anthropology: The CIA, the Pentagon, and the Growth of Dual Use Anthropology* (Durham, NC: Duke University Press, 2016), especially chap. 10.

17. Bruce C. Allnut, *Marine Combined Action Capabilities: The Vietnam Experience* (McLean, VA: Human Sciences Research, 1969), especially D1–D9. Although their primary mission was not to kill Vietcong, Combined Action Platoons personnel had a higher kill ratio (number of enemy dead compared to American casualties) than the Marines as a whole. See ibid., 11–12, F3–F10. Apparently, counterinsurgency is not necessarily a gentler form of warfare (as some have argued in recent years) but instead a more lethal and efficient one since it heavily relies on local knowledge.

18. Allison Abbe and Melissa Gouge, "Cultural Training for Military Personnel: Revisiting the Vietnam Era," *Military Review*, July–August 2012, pp. 9–17.

19. Key architects included Richard L. Humphreys, Richard W. Brislin, and Robert D. Campbell. Ibid., 16–17.

20. George E. Lowe, "The Camelot Affair," *Bulletin of the Atomic Scientists* 22, no. 5 (1966): 44–48.

21. Eric Wakin, *Anthropology Goes to War: Professional Ethics and Counterinsurgency in Thailand* (Madison: University of Wisconsin Press, 1998).

22. Laura Nader, "The Phantom Factor: Impact of the Cold War on Anthropology," in *The Cold War and the University: Toward an Intellectual History of the Postwar Years*, ed. Andre Schiffrin (New York: New Press, 1997), 107–146.

23. Renae Merle, "First Ears, Then Hearts and Minds," *Washington Post*, November 1, 2006, p. D1.

24. Hugh Gusterson, "The Cultural Turn in the War on Terror," in *Anthropology and Global Counterinsurgency*, ed. John D. Kelly, Beatrice Jauregui, Sean T. Mitchell, and Jeremy Walton (Chicago: University of Chicago Press), 279.

25. Davis, "Culture as a Weapon."

26. Ibid., 297.

27. Robert Scales, "Army Transformation: Implications for the Future," testimony before the U.S. House Armed Services Committee, July 15, 2004, pp. 4–5, available at http://www.au.af.mil/au/awc/awcgate/congress/04-07-15scales.pdf.

28. Montgomery McFate and Andrea Jackson, "An Organizational Solution for DoD's Cultural Knowledge Needs," *Military Review*, July–August 2005, pp. 18–21.

29. Roberto J. González, "Human Terrain: Past, Present, and Future Implications," *Anthropology Today* 24, no. 1 (2008): 21–26.

30. Scott Peterson, "US Army's Strategy in Afghanistan: Better Anthropology," *Christian Science Monitor*, September 7, 2007, available at http://www.csmonitor.com/2007/0907/p01s08-wosc.html.

31. Tom Vanden Brook, "Rep. Hunter Hammers Army on Social Science Project," *USA Today*, April 14, 2014, available at http://www.usatoday.com/story/news/nation/2014/04/14/rep-duncan-hunter-army-human-terrain-teams-social-science-iraq-afghanistan/7711767. Other large social science–based programs have included the Cross-Cultural Survey/Human Relations Area Files, at $4.5 million between 1941 and 1951 (worth approximately $50 million in current dollars); Project Agile, a 1960s DoD counterinsurgency research program in Thailand that made use of physicists, engineers, and social scientists at a cost of approximately $10 million annually (worth approximately $60 million in current dollars); and Project Camelot, a $6 million counterinsurgency program (worth approximately $35 million in current dollars). See Price, *Anthropological Intelligence*; Peter Braestrup, "Researchers Aid Thai Rebel Fight," *New York Times*, March 20, 1967; Irving Louis Horowitz, ed., *The Rise and Fall of Project Camelot* (Cambridge, MA: MIT Press, 1967).

32. McFate and Jackson, "An Organizational Solution," 20–21.

33. Jacob Kipp, Lester Grau, Karl Prinslow, and Don Smith, "The Human Terrain System: A CORDS for the 21st Century," *Military Review*, September–October 2006, pp. 8, 14.

34. "Anthropology and War," *The Diane Rehm Show*, October 10, 2007, available at http://thedianerehmshow.org/shows/2007-10-10/anthropologists-and-war.

35. David Rohde, "Army Enlists Anthropology in War Zones," *New York Times*, October 5, 2007, p. A1; Anne Mulrine, "Culture Warriors," *U.S. News and World Report*, December 10, 2007, 34–37; Peterson, "US Army's Strategy in Afghanistan"; Evan Goldstein, "Professors on the Battlefield," *Wall Street Journal*, August 17, 2007, available at http://online.wsj.com/articles/SB118732042794000715.

36. Matthew B. Stannard, "Montgomery McFate's Mission: Can One Anthropologist Possibly Steer the Course in Iraq?" *SFGate*, April 29, 2007, available at http://www.sfgate.com/magazine/article/Montgomery-McFate-s-Mission-Can-one-2562681.php; Louisa Kamps, "Army Brat: How Did the Child of Peace-Loving Bay Area Parents Become the New Superstar of National Security Circles?" *Elle*, April 2008, p. 309.

37. *This Week at War*, October 13, 2007, available at http://transcripts.cnn.com/TRANSCRIPTS/0710/13/tww.01.html.

38. *Charlie Rose*, December 24, 2007.

39. See Yvette Clinton, Virginia Foran-Cain, Julie Voelker McQuaid, Catherine E. Norman, and William H. Sims, *Congressionally Directed Assessment of*

the Human Terrain System (Alexandria, VA: Center for Naval Analyses, 2010), available at https://info.publicintelligence.net/CNA-HTS.pdf; National Defense University Center for Strategic Research, *Human Terrain Team Performance: An Explanation* (Washington, DC: National Defense University). The latter assessment found that "HTTs collectively were unable to make a major contribution to the counterinsurgency effort." See ibid., 6.

40. American Anthropological Association Executive Board, "Statement on the Human Terrain System Project," October 31, 2007, http://new.aaanet.org/issues/policy-advocacy/Statement-on-HTS.cfm.

41. Quoted in "SfAA Passes Motion Regarding HTS," *American Anthropological Association Blog*, May 21, 2008, available at http://blog.aaanet.org/2008/05/21/sfaa-passes-motion-regarding-hts.

42. American Anthropological Association Commission on the Engagement of Anthropology with US Security and Intelligence Communities, "Final Report on the Army's Human Terrain System Proof of Concept Program," October 14, 2009, p. 3, available at http://s3.amazonaws.com/rdcms-aaa/files/production/public/FileDownloads/pdfs/cmtes/commissions/CEAUSSIC/upload/CEAUSSIC_HTS_Final_Report.pdf.

43. Ibid., 60–61.

44. Roberto J. González, *Militarizing Culture: Essays on the Warfare State* (Walnut Creek, CA: Left Coast Press, 2010), 122.

45. Ben Connable, "All Our Eggs in a Broken Basket: How the Human Terrain System Is Undermining Sustainable Military Cultural Competence," *Military Review*, March–April 2009, p. 61.

46. Gian Gentile, "Counterinsurgency: The Graduate Level of War or Pure Hokum?" *E-International Relations*, August 3, 2013, available at http://www.e-ir.info/2013/08/03/counterinsurgency-the-graduate-level-of-war-or-pure-hokum.

47. Montgomery McFate and Steve Fondacaro, "Reflections on the Human Terrain System during the First 4 Years," *Prism* 2, no. 4 (2011): 67–68.

48. Roberto J. González, "The Rise and Fall of the Human Terrain System," *CounterPunch*, June 29, 2015, available at http://www.counterpunch.org/2015/06/29/the-rise-and-fall-of-the-human-terrain-system/.

49. Vanessa M. Gezari, "The Human Terrain System Sought to Transform the Army from Within," *Newsweek*, August 16, 2013, available at http://www.newsweek.com/2013/08/16/human-terrain-system-sought-transform-army-within-237818.html.

50. Noah Shachtman, "Pentagon's Prediction Software Didn't Spot Egypt Unrest," *Wired*, February 11, 2011, available at http://www.wired.com/2011/02/pentagon-predict-egypt-unrest.

51. "The Science of Civil War: What Makes Heroic Strife," *The Economist*, April 21, 2012, available at http://www.economist.com/node/21553006.

52. Office of Naval Research, "Broad Agency Announcement ONRBAA 11-014: Human Social Cultural and Behavioral Sciences (HSBC) Applied Research and Advanced Technology Development," 2011, available at http://www.onr.navy.mil/~/media/files/funding-announcements/baa/2011/11-014.ashx.

53. Barry Silverman, "Human Terrain Data: What Should We Do with It?" In *Proceedings of the 2007 Winter Simulation Conference*, ed. S. G. Henderson,

B. Biller, M.-H. Hsieh, J. Shortle, J. D. Tew, and R. R. Barton, 260–265 (Washington, DC: IEEE, 2007), available at http://www.informs-sim.org/wsc07papers/029.pdf.

54. Harry Goldstein, "Modeling Terrorists," *IEEE Spectrum* 43, no. 9 (2006): 30.

55. Ibid., 28.

56. University of Pennsylvania Ackoff Collaboratory for Advancement of the Systems Approach, "Overview," available at http://www.acasa.upenn.edu/nonKin/nonkin-description.htm (accessed October 29, 2014).

57. Quoted in Goldstein, "Modeling Terrorists," 31.

58. Sandra I. Erwin, "Mathematical Models: The Latest Weapons against Urban Insurgencies," *National Defense*, December 2007, available at http://www.nationaldefensemagazine.org/archive/2007/December/Pages/Mathematical2412.aspx.

59. Glenn Taylor and Ed Sims, "Developing Believable Interactive Cultural Characters for Cross-Cultural Training," in *Online Communities and Social Computing*, ed. A. Ant Ozok and Panayiotis Zaphiris, 282–291 (Berlin: Springer, 2009), 285.

60. Ibid.

61. Alelo, "Solutions," http://www.alelo.com/solutions (accessed November 5, 2014).

62. Noah Shachtman, "Air Force's Top Brain Wants a 'Social Radar' to 'See into Hearts and Minds,'" *Wired*, January 19, 2012, available at http://www.wired.com/2012/01/social-radar-sees-minds. An even stranger multimillion-dollar project was initiated by the Army Research Office several years ago to explore the possibility of using brain waves to communicate. See Donna Miles, "Army Research Grant to Explore Communication through Brain Waves," *DoD News*, September 9, 2008, available at http://archive.defense.gov/news/newsarticle.aspx?id=51091.

63. Lockheed Martin, "Integrated Crisis Early Warning System: Overview," available at http://www.lockheedmartin.com/us/products/W-ICEWS.html (accessed October 2, 2014).

64. Shachtman, "Pentagon's Prediction Software."

65. Marcus Weisgerber, "DARPA Chief: High Cost of Weapons Threatens Security," *Defense News*, May 17, 2014, available at http://www.defensenews.com/article/20140517/DEFREG02/305170023/DoD-Research-Chief-High-Cost-Weapons-Threatens-Security. See also Amir Iliaifar, "U.S. Military Developing High-Tech 'Smart' Underwear for Soldiers," *Yahoo! News*, January 26, 2012, available at http://news.yahoo.com/u-military-developing-high-tech-smart-underwear-soldiers-003417351.html.

66. Noah Shachtman, "Pentagon Forecast: Cloudy, 80% Chance of Riots," *Wired*, November 9, 2007, available at http://www.wired.com/2007/11/lockheed-peers.

67. Sharon Weinberger, "Can Social Scientists Win the War on Terrorism?" *Wired*, August 8, 2007, available at http://www.wired.com/2007/08/in-short-what-a.

68. Robert W. Cone, "NTC: The Changing National Training Center," *Military Review*, May–June 2006, 70–79.

69. Philip K. Dick, *The Minority Report* (1956; repr., New York: Pantheon, 2002).

70. C. Wright Mills, *The Sociological Imagination* (Oxford: Oxford University Press, 1959), 114. The rise to power of David Petraeus and his associates might indicate that at least some social scientists did become "generals of history."

71. Robert Albro, George Marcus, Laura A. McNamara, and Monica Schoch-Spana, *Anthropologists in the SecurityScape: Ethics, Practice, and Professional Identity* (Walnut Creek, CA: Left Coast Press, 2011), 24, 115.

72. Amy Alrich, *Framing the Cultural Landscape: Phase I Findings* (Alexandria, VA: Institute for Defense Analyses, 2008). Before the US-led invasions of Iraq and Afghanistan, a handful of anthropologists could be found at the military's elite universities, but their numbers were always quite small.

73. U.S. Marine Corps, "About CAOCL," http://www.mcu.usmc.mil/caocl/SitePages/overview.aspx.

74. Stephen J. Fallon, "Some Recent Approaches to Cultural Intelligence Gathering," *Small Wars Journal*, February 23, 2014, available at http://smallwarsjournal.com/printpdf/15296. See also U.S. Marine Corps Intelligence Activity, *Cultural Intelligence Indicators Guide* (Quantico, VA: Marine Corps Intelligence Activity, 2009), available at https://info.publicintelligence.net/MCIA-CIIG.pdf.

75. U.S. Marine Corps Intelligence Activity, *Cultural Intelligence Indicators Guide*.

76. Barak Salmoni and Paula Holmes-Eber, *Operational Culture for the Warfighter* (Quantico, VA: Marine Corps University Press, 2008), 44.

77. Ibid., 6, 33–35. Perhaps predictably, the textbook also tends to remove the U.S. military from cultural frames of reference. For example, in a discussion of "Iraqi cultural changes since 2003," the authors give no indication that the United States–led invasion and occupation might have triggered "rising inter-communal violence" or ethnic cleansing. Instead, the authors avoid delving into reasons for the rising violence: "The ethno-sectarian map of Iraq has changed, rendering earlier 'human terrain' maps obsolete. . . . Iraqi culture has changed over the course of OIF [Operation Iraqi Freedom]: in terms of body language, verbal language, dress, and demographics." Ibid., 43.

78. Elizabeth Redden, "Toward a 'Cross-Culturally Competent' Air Force," *Inside Higher Ed*, January 9, 2009, available at https://www.insidehighered.com/news/2009/01/09/air.

79. Quoted in Davis, "Culture as a Weapon."

80. U.S. Army TRADOC Culture Center, "Army Leader Exchange: Culture: We Build Adaptive Leaders," October 7, 2014, available at https://www.youtube.com/watch?v=fCB9--O1KTc&list=PLimMCO8zeiZHMw4_oxlxU6xd9mvzrAIKD.

81. General Dynamics, "Sr. Specialist, Training and Development, TRADOC Culture Center," Getting Hired, job 41717961, posted October 5, 2012, available at http://www.gettinghired.com/JobDetails.aspx?id=1576779&pnl=; ProSol Associates, "Curriculum Developer," available at https://www.smartrecruiters.com/ProSOLAssociates/79716611-curriculum-developer-vacancy-21jun2013- (accessed September 12, 2014).

82. Neil B. Carey, Edward J. Schmitz, Zachary T. Miller, and Sara M. Russell, *Assessing the Impact of and Needs for Navy Language, Regional Expertise, and Culture Training* (Washington, DC: Center for Naval Analyses, 2012), 3–4.

83. Paula Holmes-Eber, *Culture in Conflict: Irregular Warfare, Culture Policy, and the Marine Corps* (Stanford, CA: Stanford University Press, 2014), 3, 5, 6.

84. Montgomery McFate, quoted in David Rohde, "Army Enlists Anthropology in War Zones," *New York Times*, October 5, 2007, p. A1; Paula Holmes-Eber, "Teaching Culture at Marine Corps University," in *Anthropologists in the SecurityScape*, ed. Robert Albro, George Marcus, Laura A. McNamara, and Monica Schoch-Spana, 140.

85. Price, *Cold War Anthropology*, 497–498.

86. Tyler E. Boudreau, *Packing Inferno: The Unmaking of a Marine* (Port Townsend, WA: Feral House, 2008). Although National Public Radio broadcast literally dozens of programs featuring the U.S. Army and Marine Corps *Counterinsurgency Field Manual*, HTS's social scientists, and the military's in-house anthropologists, it has only once mentioned Boudreau's work—four years after his book's publication. I am grateful to Jeff Sluka for informing me about *Packing Inferno*.

87. Ibid., 32.

88. Ibid., 79.

89. Ibid., 10.

90. David B. Edwards, "Counterinsurgency as Cultural System," *Small Wars Journal*, December 27, 2010, p. 4.

91. David H. Price, *Weaponizing Anthropology: Social Science in Service of the Militarized State* (Oakland, CA: AK Press), 188.

92. Edwards, "Counterinsurgency as Cultural System," 16–17.

93. See *Human Terrain: War Becomes Academic*, directed by James Der Derian, David Udris, and Michael Udris (Oley, PA: Bullfrog Films, 2010).

94. Hugh Gusterson, quoted in ibid.

95. Franz Boas, "Why German-Americans Blame America," *New York Times*, January 8, 1916.

96. Price, *Anthropological Intelligence*, 139–150, 188–189.

Conclusion

Postnational Warfare

DAVID JACOBSON

In discussions on the transformation of war in the late twentieth and early twenty-first centuries, a recurrent theme is the notion of "new wars." The new wars thesis asserts, inter alia, that the character of war has changed such that conventional, interstate warfare is declining and nonstate, armed conflict and state-building operations are, conversely, more prominent.[1] That argument is, at least in part, aptly rebutted in a number of the chapters in this book. But in what ways *has* warfare changed now that we have moved unequivocally into the twenty-first century?

I argue in this Conclusion that we are moving into an age of *postnational* conflict—in a number of respects. It is not that state wars or belligerence has disappeared—we see otherwise in, for example, Russian and Chinese behavior. What is new is that, first, counterinsurgency and state-building operations are, in the main, taking place under postnational and multinational rubrics. These *civilizing wars* are often rooted in human rights and humanitarian law, not nationalism as such. Western states and their close allies justify their actions—from strategic rationales to tactical rules of engagement—in such human rights and humanitarian terms, with accountability to international bodies from the United Nations to the International

Criminal Court. The civilizing process—creating civil societies where the state has a monopoly on the means of violence—can be found in historical, colonial endeavors but were then decidedly national in flavor. Even the language of "civilized" nations fighting "barbarian" groups like the Islamic State of Iraq and Syria (ISIS) has reappeared, but under this different rubric.

This postnational shift is also reflected in the changing sociology of the military, especially in the West. As John Torpey and Saskia Hooiveld bring out in Chapter 1, the ideal of the citizen-soldier is declining in the world's more affluent parts. The idea that soldiering was an essential part of citizenship begins to emerge from the American Revolution and reaches its apogee in the world wars of the twentieth century, but in recent decades it has been in decline. Conscription has largely disappeared in the West. Many argue that the link between citizenship and soldiering has been severed. Some argue that military service has become in essence a job rather than a sacred duty. Further, the postnational emphasis on human rights is reflected in demands that the military reflect norms of social and political equality, with the increasing inclusion of women and gays a particular concern.

The adversaries of states are themselves postnational, especially when it comes to Islamist militants. It is not only that they are non-state actors but that they aspire to postnational political forms—and in the case of ISIS creating a caliphate has gone beyond the aspirational to the operational (although they may claim that a caliphate is actually prenational, reviving what they claim is the form of the founding, seventh-century Golden Age of Islam). These militant endeavors might be characterized as *decivilizing* movements, consciously trying to destroy the norms of global society, and especially of the West, such as human rights, religious equality, democracy, and women's empowerment. The Islamic State and Boko Haram glory in the brutal theater of crucifixion, beheadings, sex slaves, killing of Christians and other minorities, and gays thrown from tall buildings, all to make their point. Any notion of civil society that cuts across sectarian and tribal lines is undermined with graphic violence and incitement.

Ian Roxborough, in Chapter 3, argues that "theories of 'new wars' are right in asserting that, as conventional warfare between advanced states recedes into the background, the predominant form of armed conflict will be counterinsurgency and state-building operations." However, Roxborough continues, theories of "new wars" are mistaken in suggesting that such wars are novel and offers up as case studies American interventions in the Philippines and in Vietnam. Rob Johnson, in Chapter 2, proffers that, while the *character* of war is ever changing, the *nature* of war has not changed. He argues that all wars have been "characterized by the dynamic tension between the state and the people." Furthermore, attempts by states—mostly Western states—to stabilize failed or weak states with military force have been especially difficult when confronted by what Johnson calls clan systems of governance, undergirded by "irregular armed groups." This dynamic of militarily impressive states being flummoxed by clan societies or armed uprising has a deep history and for the West has antecedents well into the past.

These arguments that reject the essential underpinnings of the new wars thesis are compelling. The question remains, however, in what ways has war been transformed? After all, not *all* wars have changed, as Russia's invasion of Crimea and subsequent subterfuges in eastern Ukraine clearly demonstrate. Secretary of State John Kerry revealed that he believed that the character of war had changed—or even that war was increasingly a relic of the past—when he rather naively protested, after the Russian incursion into Crimea, "You just don't in the 21st century behave in 19th-century fashion by invading another country on completely trumped-up pretext."[2] The comment is telling, however, in its suggestion of how anachronistic the idea of seizing territory by military force has come to seem in the globe's more affluent parts.

The key differences are best revealed, I believe, when we compare the postwar decolonization struggles through the 1970s to the contemporary wars that have been waged, on one level or another, directly and indirectly, against militant Islamist movements. The decolonization struggles included a multitude of insurgencies and popular uprisings, from the Mau Mau fighting the British in Kenya, the Mozambique Liberation Front and the People's Movement for the Liberation of Angola in their struggles for independence from Portugal, the violence against the Dutch that led to a free Indonesia, or the Battle of Algiers, which traumatized France but resulted in independence for Algeria. We could even go back earlier, for example, to the 1911 republican revolution led by Sun Yat-sen in China.

The parallel, as Rob Johnson points out, lies in elements of clan politics. But the differences are even more striking. Conflicts such as those noted above are contained within the zones of contention—that is, within the boundaries of the colonial territory itself (say, Kenya). The conflicts were not globalized or even very often taken to the colonial metropole, be it London, Paris, or elsewhere. One could attribute this to the technologies of war of the time, or to the limited ability to travel, or to the inability of small diasporas to extend the wars to Western metropoles. But the capacity to take such violence into the metropole existed via the many students from the colonies who studied in such capitals as London or Paris (and there were violent protests by Algerians living in France spilling over during the Algerian independence struggle).

The key difference between those conflicts and those of our own day is, in fact, ideological. If we examine the ideological grounding of the anticolonial, national liberation leadership, they used the language and vocabulary of the West. They sought an equal place under the sun: their own nation-state, national self-determination, and independence to be a sovereign state and people in each of the colonies. The colonies, now nation-states, aspired to join, mutatis mutandis, the Westphalian system. Initially, at least, they did not even challenge the often arbitrary boundaries of states that had been determined by the colonial powers.

Later on, struggles within postcolonial states to create new rump states, frequently based on tribal differences, were generally rejected as illegitimate by Western states—for example, in the cases of the National Union for the Total Independence of Angola (UNITA) movement in southern Angola or Biafra in Nigeria. Bangladesh, carved out of East Pakistan, is a successful exception. Even where two nations or peoples struggled over the same land—say, the Indo-Pakistani struggle over Kashmir or that of the Israelis and Palestinians—these struggles were based on the same Western principles of nationalism and independence. In the Palestinian case, however, this does change with the rise of Islamist groups like Hamas, a point further elaborated below.

It is important to note here that the anticolonial liberation elite were largely educated in the West, in Western universities, and often at elite schools like the London School of Economics or the Sorbonne. Jomo Kenyatta, for example, who would lead Kenya to independence and serve as the country's first president, studied anthropology at LSE. These leaders absorbed and were drawn to Western political concepts and to the lure of science and industrialization as they developed in the West. They by and large rejected traditional clans and their leadership, except perhaps on a tactical level.[3] Western ideas, then, were not simply *exported* by the West to its colonies but were in fact also eagerly *imported* by the liberation leadership.

Unfortunately, the promise of democracy was extensively supplanted by the rule of brute force, as these same leaders were desperate to hold on to power in divided societies. In the case of Sun Yat-sen (who was educated at a then relatively unknown university, the University of Hawaii), the warlords and the communists under Mao overwhelmed his idealism and democratic and nationalist aspirations. There are a handful of exceptions; Botswana, for example, managed to remain a democracy from independence in 1966 to the present day. Others did not get much of a chance to create a democratic order: Patrice Lumumba served as the first democratically elected prime minister of the newly independent Democratic Republic of the Congo (DRC) for a mere twelve weeks before he was assassinated by provincial secessionists backed by Western interests.

One could argue that the influence of Marxism across a swath of the anticolonial movement indicated an anticolonial thread that was categorically un-Western. But this is only half right at best. In a sense, Marxism and, even more so, Marxism-Leninism allowed members of anticolonial liberation movements to reconcile two divergent tendencies. They were drawn to the Western promise of science, technology, and industrial progress. Yet they had witnessed the humiliation of their countries under colonial rule and in many cases had personally suffered slights and racism. Marxism provided a narrative of history—Marx's famous five historical and purportedly universal stages of primitive hunting band, ancient societies, feudal, capitalist, and finally socialist—that implied that industrial development was not a function of Western rule but intrinsic to every society. Lenin cemented this narrative in his book *Imperialism: The Highest*

Stage of Capitalism by suggesting that the struggle was not so much internal but international class struggle, ascribing to the decolonizing countries the role of the liberating proletariat.[4] In other words, Marxism-Leninism provided a bridge for a satisfying story of self-emancipation into what was, in practice, a Western-dominated world of modern states. The colonized became the agents of history rather than its victims.

The important point about decolonization movements is this: once independence was achieved, the violence and struggle could end. With shared vocabulary like "national self-determination," one could draw a line in the sand. Once the British left Kenya, for example, the war was over and the Mau Mau disbanded. Furthermore, though clan and tribal politics played a key role during and after liberation (and still to this day), the leadership aspired, in principle at least, to a civil, republican society. And though they might at various times have leaned more to the West or more to Moscow, in principle the anticolonial leadership sought to join the society of states and not to destroy it in toto.

Contemporary Islamist militant groups have brought three major changes into warfare,[5] in ways that are not captured in either the new wars thesis or in critiques of it. These changes become clearer when these groups are compared to the protagonists of the anticolonial struggles.

First, militant Islamist groups have invoked the notion of the caliphate, a theocratic order lacking the national boundaries associated with the Westphalian order. Even in Islamist movements that do not declare a caliphate immediately, the vastly different ideological approach is reflected in, among other things, demands for the implementation of more extreme versions of sharia, including *hudud* ("crimes against God"). The Islamic State's sense that they are bringing on the apocalypse reflects this opposition to the Westphalian order. The savagery of groups like ISIS or Boko Haram is not merely about reverting to seventh-century norms of war; it is an exhibitionist rejection of international norms of war and humanitarian principles.

Second, they have also both incorporated *and* transcended clan politics and war in a novel way. In other words, key tribal values, such as patriarchy, honor, grievance, and feuds were absorbed by Islamist groups but now expressed not as, say, intertribal concerns, based on kinship, but in global terms. The enemies could now be more abstract—from supposedly apostate Muslim governments to the United States and the West, "Crusaders," and Jews.[6]

Third, they have created a conflict that is simultaneously national—for example, the conflicts in Iraq, Syria, Afghanistan, and Somalia—and global. Now the United States, Europe, Russia, and China, among other areas, are part of the theater of operations for Islamists. We are witnessing, to use the sociologist Roland Robertson's term, the "glocalization" of war.[7]

While postwar, anticolonial movements on occasion projected violence into the metropole—although such violence was marginal relative to the level of violence in the home country—it was strictly of a tactical nature. There was no pretense of bringing the colonists under the umbrella of their sovereignty or shaping law to conform with something like sharia. The Jordanian leader, King Abdullah, suggested, in a CNN interview by Fareed Zakaria on March 1, 2015, that we are seeing the emergence of World War III.[8]

We can point to a number of drivers of this conflict: for example, the globalization of communications, including the Internet, has facilitated the spread of global ideologies. As Derek Harvey, a former senior intelligence officer in the Defense Intelligence Agency and senior advisor to General David Petraeus in Iraq, has pointed out, we are also witnessing the democratization of the means of violence (and in some respects, the means of intelligence).[9] Such global developments marry with local or regional challenges—for example, the youth bulge or the production of students in schools driven often purely by religious learning, unequipped for the modern, globalized world. New elites emerge, resentful of their marginal status in an essentially Western-defined world. (In my research in Nigeria in 2015, I found that it is the better off, not necessarily the poor or even the most pious, who are more likely to support movements such as Boko Haram in Nigeria.)

But those are structural factors that make such movements possible—a necessary but not sufficient sociology. The larger question is what are the underlying ideological factors that make these Islamist calls for violent jihad so attractive? What drives the underlying agency that structural changes facilitate?

The new wars argument, notably that of Mary Kaldor, posits that wars are now driven by identity, not territory; that they involve the use of terrorism and guerrilla warfare; that most victims are civilians rather than uniformed soldiers; and that ethnic divisions are exacerbated and state integrity is brought into question. The argument contains partial truths but at the same time misses more nuanced developments with an underlying logic that are not as centrifugal as Kaldor and others suggest.

One is tempted to use the language of G.W.F. Hegel—that of the dialectic—to make sense of these developments. The dialectic involves a chain of thesis-antithesis-synthesis. The synthesis is not simply additive but rather indicates a transcendent new form while still capturing elements of the old forms. What is extraordinary about Islamist movements is that they both contain old forms of the tribes and clans yet shape them into the form of a new ideological movement that is ultimately global. In this regard, they go beyond the state-clan clash that Rob Johnson points to—and explains why conflict has become "glocalized."

Tribalism has played an extraordinary role in the context of the struggle with militant Islamists—from Afghanistan and Pakistan to Yemen, the Horn of Africa, and across North Africa. Indeed, tribalism, including in diasporas, constitutes an essential element of the social landscape of militant Islamist movements. Understanding the links between tribalism and global Islamist movements—the tribal–Islamist nexus—is essential to unlocking the sociological and political dynamics of the broader conflict. One commentator, alluding in 2010 to the wars involving Western powers in Afghanistan and Iraq—and, less overtly, in Yemen, Somalia, North Africa, and elsewhere—suggested that the "Global War on Terror" was actually a "Global War on Tribes."[10]

This nexus between Islamism and tribalism is a part of ideological struggles from the Middle East to Western Europe. Wherever migrant populations have an ancestry in societies that are highly tribal, an element will, it turns out, be more prone to religiously motivated violence. Islamic identification in isolation, independent of these tribal qualities, is generally not sufficient in explaining this violence. Even urbanized societies and diaspora communities can maintain qualities of tribalism. In fact, the role of tribal regions in this conflict with militants is not coincidental, and it is tightly interwoven with the extremely rigid patriarchy of these regions.[11]

Militant Islamist groups are most likely to develop in tribal patriarchal environments. This suggests that the intersection of tribal patriarchy and Islam is where these militant groups are best nurtured. They may still constitute a relatively small percentage of the population as a whole. But militant Islamists are more likely to prosper in such a social environment. And it is Islamist movements that in turn are more likely to export violence internationally. High levels of patriarchal tribalism are interwoven with Islamist movements—but Islam independent of patriarchal tribalism has a much less significant, indeed a negative, relationship with religiously motivated violence.[12]

That said, the normally kinship-based, geographically contained character of tribal violence takes on an ideological and globalized form when it occurs among Muslims (in an Islamist form). Tribal conflicts as a whole tend to be fairly insular: if the fight is based on kinship issues, the quarrels are on the level of families and clans. In more extreme cases—witness Rwanda and the Democratic Republic of Congo—the bloodletting will take on a broader reach. Such conflict, however, rarely spills over beyond one or two countries. In other words, the violence does not become global or transnational. It is not sufficient simply to look at tribes and tribal patriarchy and tribalism per se. It is this linkage of tribalism and Islamism that is so problematic for Western security interests.[13]

The new wars thesis does not capture the full complexity and nuance of changes in warfare. It is not just that we have nonstate actors or clan politics or the use of terrorism and the like. The resistance of nonstate groups to Western states and their influence is well beyond anything we saw in the decolonization struggles after World War II. Overlaid on the more recent combat is the growing role of what Derek Harvey calls the "human dimension" of war.[14] Roberto González extensively criticizes the use of the social sciences by the U.S. military in Chapter 7, arguing that this is mostly window dressing that cannot mask more brutal realities. Notwithstanding González's critique, attention to the "human dimension" reflects and reinforces the humanitarian and human rights constraints on Western militaries—which C. Anthony Pfaff discusses in Chapter 4, on the paradoxes of the recent rules of engagement. (While these constraints were not altogether absent in the past, neither were they as robust as they are today.) New technologies, notably drones and biometrics, provide the option—some would say the illusion—of cleaner, apparently more discriminating uses of violence (see Chapters 5 and 6).

The brilliant sociologist Norbert Elias was recognized very late in his life for his work on the "civilizing process."[15] He illustrated the interlinked importance of the state monopoly of violence with the growing self-restraint essential for a civil society and for republican forms of government. The emergence of civility led to more psychological and less physical modes of deportment as the legitimate use of violence was banished from polite society and monopolized by states. That observation is now well known. But we still do not understand the full import of Elias's work—namely, that the Westphalian system was not only international, or interstate, but depended on the recasting of domestic politics as well. The civilizing process was international in its import as well as domestic.

In the rise of Islamist movements—represented in their most extreme form by ISIS and its allies in groups like Boko Haram—we see *decivilizing* movements. Their warfare has collapsed the categories of international and domestic (except in narrow tactical terms); they have equally collapsed the distinction between public and private. In the words of the Algerian writer Kamel Daoud, "Islamism takes care of you completely: your body, your sexuality, your life, your relations with others." He adds, suggesting an ideological vacuum in the Middle East outside Islamism, "On the other side, there is nothing. This is the philosophical disaster of the Arab world."[16]

For ISIS, and others like Al Qaeda and in somewhat moderated forms groups such as the Muslim Brotherhood, the implementation of sharia involves a divine code that does not (in the Islamist iteration) recognize fundamental divisions of public and private, state and society. It is seamless—and that seamlessness is writ global through the caliphate, which does not recognize international borders as having any legal or sacred writ. Indeed, here sharia expresses the social and political seamlessness both vertically in society and horizontally across the globe.

Borders of the caliphate are temporal facts, driven by circumstances of war or of *hudna*—a temporary lull, at best an armistice—that separate truly Muslim states (that is, Islamist theocracies) from supposedly apostate Muslim states and from non-Muslim states and peoples (*dhimmi*). It is the division between the realm of Islam (*dar al-Islam*) and other polities that are, tellingly, in the realm of war (*dar al-harb*). Sayyid Qutb, the father of modern Islamism, was a leading proponent of this vision in his most famous text, *Milestones*.[17] Tactical differences certainly exist, but the principles, collapsing public and private, domestic and international, cut across the Islamist movements.

The Islamist uprising attacks the bodies politic of the West—indeed, of the world—challenging their constitutive foundations. Simple distinctions between law enforcement and war itself will not suffice either. This is not a clash of civilizations, for myriad reasons, not least that the Muslim world is itself ideologically split over the Islamist presence and that the primary victims of Islamist depredations are Muslims.

I begin this concluding chapter by noting critics of the new wars thesis who agree that conventional state-to-state war will recede into the background and the predominant form of armed conflict will be counterinsurgency and state-building operations, but the critics also argue that theories of new wars are mistaken in suggesting that such counterinsurgency wars are that new. What *is* new is that counterinsurgency and state-building operations are no longer *national* in the way they were for, say, the British in Malaysia. While, oddly, the language of the colonial past has reappeared—"civilized" nations fighting "barbarian" groups like ISIS—these civilizing wars are taking place under a distinctly different, postnational rubric—namely, that of human rights and humanitarian law. A new accountability to international bodies, from the UN to the International Criminal Court, colors the actions of Western states and some of their close allies.

The contrast between this new postnational rendering of civilization against barbarity is starkly illustrated, for example, in the attitudes to women and sexuality: in the West we find a growing emphasis on the inclusion of women and gays in the military. Among Islamist militants, keeping women in traditional, patriarchal subordination is a core ideological goal, with a central grievance against the West regarding the alleged corruption of women in the Muslim world. Women are recruited by ISIS to marry and support jihadis as well as police other women, while women of minorities are subject to particular humiliation and brutality as sex slaves, rape, and forced marriage. Gays, or allegedly gay men, are dropped from buildings to their deaths.

National interests of treasure and power remain—and Russian and Chinese behavior on the international stage should certainly give one pause in this respect. In principle, however, states are now accountable to norms of human rights rather than simply to sovereign prerogative. Western states generally stand closely aligned in their military interventions. Western intelligence services closely cooperate. American transport planes, for example, ferry French troops to Mali. Western interests are jointly wrapped up in the civilizing process, which, ironically, limits those same states in how they conduct themselves in war. (The brutal counterinsurgencies of the British in Malaysia or in French Algeria, to name just two examples, would not be acceptable today.) Those limits are amply illustrated in the discussions in this book, notably in C. Anthony Pfaff's discussion of the contemporary rules of engagement (Chapter 4).

"New wars" indeed, but very different from what Kaldor and others pictured. Yet the situation is in flux. We are still slowly and unsteadily working our way to defining a strategy (and tactics) to match the scope and complexity of this new, postnational battlespace.

Notes

1. Mary Kaldor, *New and Old Wars: Organized Violence in a Global Era* (Stanford, CA: Stanford University Press, 1999).

2. Will Dunham, "Kerry Condemns Russia's 'Incredible Act of Aggression' in Ukraine," *Reuters*, March 2, 2014, available at http://www.reuters.com/article/2014/03/02/us-ukraine-crisis-usa-kerry-idUSBREA210DG20140302.

3. I say "by and large": Kenyatta, for example, made a defense of female circumcision, or what many consider female genital mutilation, but he rejected traditional tribal leadership as a basis for modern rule, even while drawing on his Kikuyu tribal base for political support.

4. V. I. Lenin, *Imperialism: The Highest Stage of Capitalism* (New York: International, 1939).

5. Graeme Wood, "What ISIS Really Wants," *Atlantic Monthly*, March 2015, available at http://www.theatlantic.com/magazine/archive/2015/03/what-isis-really-wants/384980.

6. Sayyid Qutb (1906–1966), one of the most notable ideologists of modern Islamist thought, resolves the tension between a universal, monotheistic faith and

the local tribal patriarchy by, first, seeking a pure Islam—free of the "taints" of locality, culture, tribe, or nation—while taking tribal patriarchy as the foundation of a global Islam. This made sense for Qutb's reading of the Koran. The Prophet himself sought to transcend the warring, fractious tribes of Arabia (although Qutb does note Muhammad's own, noble tribal lineage). See the discussion in David Jacobson, *Of Virgins and Martyrs: Women and Sexuality in Global Conflict* (Baltimore: Johns Hopkins University Press, 2014), chap. 7.

7. Roland Robertson, "Glocalization: Time-Space and Homogeneity-Heterogeneity," in *Global Modernities*, ed. Mike Featherstone, Scott Lash, and Roland Robertson (London: Sage, 1995), 25–43.

8. Fareed Zakaria, "Jordan's King Abdullah: 'This Is a Third World War by Other Means,'" *CNN*, March 1, 2015, available at http://cnnpressroom.blogs.cnn.com/2015/03/01/jordans-king-abdullah-this-is-a-third-world-war-by-other-means.

9. Derek Harvey, personal communication, March 1, 2015.

10. Zoltan Grossman, "The Global War on Tribes," *Counterpunch*, April 13, 2010, available at http://www.counterpunch.org/2010/04/13/the-global-war-on-tribes.

11. Jacobson, *Of Virgins and Martyrs.*

12. Ibid.

13. Ibid.

14. Derek Harvey, personal communication, November 2, 2014.

15. See Norbert Elias, *The Civilizing Process* (Oxford: Blackwell, 1994).

16. Quoted in Elisabeth Zerofsky, "An Algerian Rebuke to 'The Stranger,'" *New Yorker*, March 13, 2015, available at http://www.newyorker.com/news/news-desk/an-algerian-in-paris-kamel-daoud.

17. Denis MacEoin, "Tactical Hudna and Islamist Intolerance," *Middle East Quarterly*, Summer 2008, pp. 39–48.

Contributors

Ariel Colonomos is senior research fellow at the National Center for Scientific Research (Centre d'Etudes et de Recherches Internationales; CNRS) and research professor at Sciences Po in Paris, where he teaches courses on international relations theory and the ethics of war. He has published in the areas of international relations, the ethics of war, and political sociology. His most recent books include *Moralizing International Relations: Called to Account* (2008), *The Gamble of War: On Justifying Preventive War* (2013), and *Selling the Future: The Perils of Global Predictions* (2015).

Roberto J. González is professor of anthropology at San Jose State University, California. He has published several books, including *Zapotec Science: Farming and Food in the Northern Sierra of Oaxaca* (2001), *Anthropologists in the Public Sphere: Speaking Out on War, Peace, and American Power* (2004), *American Counterinsurgency: Human Science and the Human Terrain* (2009), and *Militarizing Culture: Essays on the Warfare State* (2010). His most recent book is the coedited volume (with Rachael Stryker) *Up, Down, and Sideways: Anthropologists Trace the Pathways of Power* (2014). He also coproduced (with Laura Nader) the documentary film *Losing Knowledge: 50 Years of Change* (2013).

Travis R. Hall is a professional lecturer at the School of International Service at American University in Washington, D.C.

Saskia Hooiveld is a doctoral fellow in the Ph.D. Program in Sociology at the Graduate Center, City University of New York. Her dissertation, "Secrecy on

Display: Claims-Making in Exile Media Organizations," investigates the forms of communication created by exiles as they seek to continue to engage in politics in their countries of origin. She teaches courses in sociology at Queens College of the City University of New York.

David Jacobson is founding director of the Global Initiative on Civil Society and Conflict and professor of sociology at the University of South Florida. His most recent book is *Of Virgins and Martyrs: Women and Sexuality in Global Conflict* (2014). Previously he authored *Rights across Borders: Immigration and the Decline of Citizenship* (1997) and *Place and Belonging in America* (2001). He has directed research projects in West Africa, Western Europe, and the United States and conducted surveys in eight countries.

Rob Johnson is the director of the Changing Character of War (CCW) research program at Oxford University. His specialism is in the history of war and the issues of future projection. His research is also concerned with regional studies of insurgency and state fragility. He has published *The Iran-Iraq War* (2010), *The Afghan Way of War* (2011), and *At the End of Military Intervention* (2014).

C. Anthony Pfaff is a colonel and foreign area officer serving as the senior military and Army advisor on the Policy Planning Staff at the Department of State. Before this assignment he was the defense attaché in Baghdad, senior military advisor to the Civilian Police Assistance Team in Iraq, and deputy chief of intelligence for a joint counterterrorism task force in Iraq. Originally an infantry officer, he served in the Eighty-Second Airborne Division, in the First Armored Division, and on the faculty at West Point, where he taught philosophy.

Ian Roxborough is professor of sociology and history at Stony Brook University. He is coeditor (with L. Bethell) of *Latin America between the Second World War and the Cold War* (1992), author of *Unions and Politics in Mexico* (1984) and *Theories of Underdevelopment* (1979), and coauthor (with P. O'Brien and J. Roddick) of *Chile: State and Revolution* (1977).

John Torpey is professor of sociology and history at the Graduate Center, City University of New York, and director of its Ralph Bunche Institute for International Studies. He is the author or editor of numerous books, including *Old Europe, New Europe, Core Europe: Transatlantic Relations after the Iraq War* (edited with Daniel Levy and Max Pensky; 2005), *Making Whole What Has Been Smashed: On Reparations Politics* (2006), and *Legal Integration of Islam: A Transatlantic Comparison* (with Christian Joppke; 2013).

Index